ORGANIC REACTIONS IN AQUEOUS MEDIA

ORGANIC REACTIONS IN AQUEOUS MEDIA

CHAO-JUN LI
Tulane University
New Orleans, Louisiana
USA

TAK-HANG CHAN
McGill University
Montreal, Quebec
Canada

A Wiley-Interscience Publication
JOHN WILEY & SONS, INC.
New York / Chichester / Weinheim / Brisbane / Singapore / Toronto

Li, Chao-Jun, 1963–
    Organic reactions in aqueous media / Chao-Jun Li and Tak-Hang
Chan.
        p.   cm.
    "A Wiley-Interscience publication."
    Includes index.
    ISBN 0-471-16395-3 (alk. paper)
    1. Chemistry, Organic. 2. Solution (Chemistry) 3. Chemical
reactions.   I. Chan, Tak-Hang, 1941–   II. Title.
QD255.4.L5   1997
547′. 13422 — dc21   96–29886

Printed in the United States of America

10   9   8   7   6   5   4   3   2   1

To our wives and children

CONTENTS

PREFACE

This book covers organic reactions in aqueous media. It is arranged in seven chapters, starting with the fundamental properties of water to pericyclic reactions and then covering nucleophilic additions and substitutions; metal-mediated reactions; transition-metal-catalyzed reactions, oxidations, and reductions; and industrial applications. Other reactions, such as enzymatic reactions and phase-transfer-catalyst-catalyzed reactions, are discussed only briefly because more specialized monographs are available in the literature.

The use of water as solvent has a special effect on the reaction and this fact has been considered in the selection of most of the materials covered in this book. Instances, such as hydrolysis, in which the use of water is considered routine or in which water is used as a simple reagent have not been included. The distinction of routine vs. non-routine is less obvious for some reactions. In these cases, the selection of coverage is purely arbitrary; there is no intention to undermine the importance of those works not covered.

Finally, we would like to thank John Haberman, William T. Slaven IV, and Chris Costello for their technical assistance.

CHAO-JUN LI
New Orleans, Louisiana, USA

TAK-HANG CHAN
Montreal, Quebec, Canada

ORGANIC REACTIONS IN AQUEOUS MEDIA

CHAPTER 1

INTRODUCTION

Of all the inorganic substances acting in their own proper nature, and without assistance or combination, water is the most wonderful. If we think of it as the source of all the changefulness and beauty which we have seen in clouds; then as the instrument by which the earth we have contemplated was modeled into symmetry, and its crags chiseled into grace; then as, in the form of snow, it robes the mountains it has made, with what transcendent light which we could not have conceived if we had not seen; then as it exists in the form of torrent, in the iris which spans it, in the morning mist which rises from it, in the deep crystalline pools which mirror its hanging shore, in the broad lake and glancing river; finally, in that which is to all human minds the best emblem of unwearied, unconquerable power, the wild, various, fantastic, tameless unity of the sea; ... It is like trying to paint a soul.

—Ruskin*

Aqueous chemistry is one of the oldest forces of change in the solar system. It started less than 20 million years after the gases of the solar nebula begun to coalesce into solid objects (1). Water is also the most abundant volatile molecule in comets. On the earth, the oceans alone contain ~1.4×10^{21} kg or 320,000,000 mi^3 of water. Another 0.8×10^{21} kg is held within the rocks of the earth's crust, existing in the form of water of hydration. The human body is roughly 65% water

*Initially written by J. Ruskin, cited from W. Coles-Finch, and E. Hawks, *Water in Nature*, T. C. & E. C. Jack Ltd., London, 1933.

1

by weight (wt%). Some tissues like brain and lung are composed of nearly 80% of water (2).

Water is the basis and bearer of life. For millions of years, water had been at work to prepare the earth for the evolution of life. It is the solvent in which numerous biochemical organic reactions (and inorganic reactions) take place. All these reactions affecting the living system have inevitably occurred in aqueous media. On the other hand, the development of modern organic chemistry has been based almost exclusively on the fact that organic reactions are often carried out in organic solvents. Researchers have again focused their attention on carrying out organic reactions in water only within the last decade or so. This resurgence is due in large part to the study by Breslow on Diels–Alder reactions (3). Since then, many organic reactions that are traditionally carried out exclusively in organic solvents, such as the Barbier–Grignard-type reaction, have been successfully performed in aqueous media. Novel reactions for which the use of water as solvent is critical have also been discovered.

Why should we consider using water as a solvent for organic reactions? There are many potential advantages:

- *Cost.* Water is the cheapest solvent available on earth; using water as a solvent can make many chemical processes more economical.
- *Safety.* Many organic solvents are inflammable, potentially explosive, mutagenic, and/or carcinogenic. Water, on the other hand, is not.
- *Synthetic Efficiency.* In many organic syntheses, it may be possible to eliminate the need for protection and deprotection of functional groups, and save many steps in the synthesis process. Water soluble substrates can be used directly. This will be especially useful in carbohydrate and protein chemistry.
- *Simple Operation.* In large industrial processes, isolation of the organic products can be performed by simple phase separation. It is also easier to control the reaction temperature, since water has one of the largest heat capacities of all substances.
- *Environmental Concerns.* The use of water as solvent may alleviate the problem of pollution by organic solvents since water can be recycled readily.

Finally, compared to reactions in organic solvents, the use of water as a reaction solvent has been explored much less in organic chemistry.

There are many opportunities to develop as yet undiscovered novel synthetic methodologies.

This chapter will briefly survey some basic physical and chemical properties of water and the possible relevance of these properties to aqueous organic chemistry.

1.1 THE STRUCTURE AND FORMS OF WATER

In the 1780s, Cavendish and Lavoisier established that water is composed of hydrogen to oxygen. Gay-Lussac and Humboldt discovered, in 1805, that the ratio of hydrogen to oxygen in a water molecule is $2:1$. And, in 1842, Dumas found that the ratio of the combining weights of hydrogen and oxygen is very close to $2:16$ in the molecule (4).

Water has two σ bonds, two lone pairs of electrons on oxygen, and a bond angle of 104.5° at oxygen.

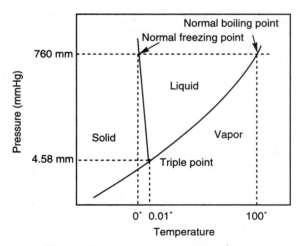

Water exists in three basic forms: vapor, liquid, and solid. The relationships between these three forms of water are described by the pressure–volume–temperature phase diagram. (Fig. 1.1).

Figure 1.1 Phase diagram for water.

The structure of water in its liquid state is very complicated and remains a topic of current research. The structure of liquid water, with its molecules interconnected by hydrogen bonds, gives rise to several anomalies when compared with other liquids (5).

In its solid state, however, the basic structural features of ordinary hexagonal ice (ice I) are well established. In this structure (Fig. 1.2), each water molecule is hydrogen-bonded to four others in nearly perfect tetrahedral coordination. This arrangement leads to an open lattice in which intermolecular cohesion is large.

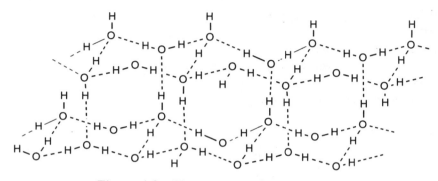

Figure 1.2   The structure of hexagonal ice.

1.2 PROPERTIES OF WATER

The principal physical properties of water are shown in Table 1.1 (6).

TABLE 1.1 Principal Physical Properties of Water

Density g/mL (3.98°C)	1.0
Melting point, °C (at 760 mmHg)	0.0
Boiling point, °C (at 760 mmHg)	100.0
Temperature at maximum density, °C	3.98
[cal/(g · °C)]	[79.71]
Heat of melting, J/g °C	333.75
Heat of vaporization, cal/(g · °C)(0°C) [J/(g · °C)]	595.4 [2260]
Specific heat, cal/(g · °C)(15°C) [J/(g · °C)]	1.00 [4.19]
Surface tension, mN/m (20°C)	72.75
Dynamic viscosity, (mN · s)/m^2 (20°C)	1.000
Specific electrical conductivity, S/m (25°C)	5.10^{-6}
Critical temperature, °C	374.0
Critical density, g/cm^3	0.322
Critical specific volume, cm^3/g	3.11
Dielectric constant (20°C)	80.20

Water is at peak density at 3.98°C; water density decreases as the temperature falls to 0°C. For this reason, ice is lighter than water and floats, which insulates the deeper water from the cold temperature and prevents it from freezing. This property has fundamental importance in nature. The density of water also decreases when the temperature exceeds 3.98°C. It reaches the same density of ice at about 70°C.

The viscosity of water also changes with temperature. Water decreases viscosity inversely with temperature because the number of hydrogen bonds binding the molecules together reduces when the temperature rises. The viscosity of water affects the movement of solutes in water and the sedimentation rate of suspended solids.

Water has the highest value for specific heat of all substances. The specific heat of water is the standard against which the values for all other substances are determined. The high value for the specific heat of water is due to the great heat capacity of the water mass. This means that rapid changes in ambient temperature result in slow changes in water temperature. Such an effect is important for aquatic organisms. It is also an important advantage in the use of water as solvent to control the temperature for both endo- and exothermic reactions, especially in large-scale industrial processes.

If the surface of a liquid is regarded as an elastic membrane, then the surface tension is the breaking force of this membrane. Water has one of the highest surface tensions of all liquids. For example, the surface tension of ethanol at 20°C is 22 mN/m, while that of water is 72.75 mN/m. The surface tension of water decreases with temperature. The presence of surface-active agents (surfactants), such as detergents, also decreases the surface tension of water.

1.3 SOLVATION

Water is a very good solvent for many substances, which is of fundamental importance in nature. The solubility of chemical substance is dependent on the temperature. The solubility of gases, such as oxygen, nitrogen, and carbon dioxide, in water usually decreases with a rise in temperature. However, there are some gases whose solubility increases with an increase in temperature. An example of such is helium. Similarly, there is variation in the relation between the solubility of solids in water and the temperature. For example, the solubility of $AgNO_3$ rapidly increases with an increase in temperature, but for NaCl, there is only a slight increase in solubility with a rise in temperature. On the other hand, when the temperature increases, there is a decrease in solubility for Li_2CO_3. The influence of the temperature on the solubility of substances is dictated by the heat of solution of the substance, which is the heat emitted or absorbed during the dissolution of one mole of a substance in one liter of water.

Metal ions in aqueous solution exist as complexes with water. The solubility of organic compounds in water depends primarily on their polarity and ability to form hydrogen bonds with water. Organic compounds with a large part (weight percentage) of polar components, such as acetic acid, dissolve in water without limit. In such cases, the polar part dominates. In contrast, soaps and detergents have a polar "end" attached to a relatively large nonpolar part of the molecule. They have limited solubility, and the molecules tend to coalesce to form micelles.

1.4 HYDROPHOBIC EFFECT

Polar compounds and compounds that ionize can dissolve readily in water. These compounds are said to be hydrophilic. In contrast to hydrophilic substances, hydrocarbons and other nonpolar substances have very low solubility in water because it is energetically more favorable for water molecules to interact with other water molecules than with nonpolar molecules. As a result, water molecules tend to exclude nonpolar substances, forcing them to associate with themselves in forming drops, thereby minimizing the contact area between water and the organic substance. This phenomenon of repulsion of nonpolar substances by water is referred to as the *hydrophobic effect* (7). The hydrophobic effect plays a critical role in biological systems. For example, it is the primary force in determining the folding patterns of proteins and the self-assembly of biological membranes.

Hydrophobic interaction, the association of a relatively nonpolar molecule or group in water with other nonpolar molecules, is due not to the mutual attraction of the nonpolar molecules but to the large cohesive energy density (CED) of water. The large cohesive energy density causes the polar water molecules surrounding the nonpolar compounds to associate with each other. The internal pressure of water, however, may also play a role in hydrophobic interaction. Table 1.2 contains the internal pressure and cohesive energy density for some common solvents at 25°C.

In thermodynamic terms, solutes can be divided into two classes. For hydrophobic solutes in dilute solution in water, the partial Gibbs

TABLE 1.2 Internal Pressures and Cohesive Energy Densities for Some Common Solvents (25°C) (9)

Solvent	CED (cal/cm^3)	Internal Pressure (cal/cm^3)
Water	550.2	41.0
Formamide	376.4	131
Methanol	208.8	70.9
Dimethyl sulfoxide	168.6	123.7
Dimethylformamide	139.2	114
Acetonitrile	139.2	96
Acetone	94.3	79.5
Benzene	83.7	88.4
Carbon tetrachloride	73.6	80.6
Diethyl ether	59.9	63.0
Hexane	52.4	57.1

free energy of solution is positive. This is because water molecules that surround a less polar molecule in solution are more restricted in their interactions with other water molecules, and these restricted water molecules are relatively immobile and ordered. However, water molecules in the bulk solvent phase are much more mobile and disordered (Fig. 1.3a). Thermodynamically, there is a net gain in the combined entropy of the solvent and the nonpolar solute when the nonpolar groups aggregate and water is freed from its ordered state surrounding the nonpolar groups (Fig. 1.3b).

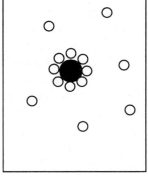

(a)   (b)

Figure 1.3   (● nonpolar solute; ○ water).

Hydrophobic interaction, however, is a relatively weak interaction. For example, the energy required to transfer a $-CH_2-$ group from a hydrophobic to an aqueous environment is ~3 kJ/mol.

Thus, the transfer of a hydrophobic molecule from a pure state to an aqueous solution is an unfavorable process because of the large decrease in entropy resulting from the reorganization of the water molecules surrounding the solutes. In contrast, the partial Gibbs free energy for dissolving a hydrophilic solute in water is negative. This is because hydrophilic solutes can bind water molecules through hydrogen bonding.

1.5 SALT EFFECT

The effect of dissolved hydrophilic electrolytes on the interaction between organic solutes and water can be described by the salting-in and salting-out effects. Dissolved electrolytes usually increase the internal pressure in water through a volume-reducing process that involves polarization and attraction of solvent molecules around the ionic species (electrostriction). For example, a 3 M aqueous solution of sodium bromide has an internal pressure of ~75 cal/cm^3, whereas the internal pressure of water at 25°C is only 41 cal/cm^3.

When the dissolved salt increases, the internal pressure of the aqueous solution—to a certain extent, the nonelectrolyte—is squeezed out (salting-out effect). On the other hand, when the dissolved salt reduces the internal pressure of the solution, more of the nonelectrolyte is able to dissolve (salting-in effect). All the electrolytes, except perchloric acid, increase the internal pressure of water and cause a salting out of organic species. For example, saturated sodium chloride is used to separate organic compounds from water.

Water's internal pressure acts on the volume of activation ($\Delta V^{\neq}$) of a reaction in the same way as an externally applied pressure does. Thus, the internal pressure of water influences the rates of nonpolar reactions in water in the same direction as external pressures. Nonpolar reactions with a negative volume of activation will thus be accelerated by the internal pressure of water, whereas nonpolar reactions with a positive volume of activation will be slowed by the internal pressure. For example, at 20°C the rate of Diels–Alder reaction between cyclopentadiene and butenone in a 4.86 M LiCl solution is about twice that of the reaction in water alone (3):

(1.1)

Solvent	$k_2 \times 10^5$ (M^{-1}s^{-1})
H_2O	4,400 $\pm$ 70
H_2O + LiCl (4.86 M)	10,800

1.6 WATER UNDER EXTREME CONDITIONS

Ordinary water behaves very differently under high temperature and high pressure (9). Early studies of aqueous solutions under high pressure showed a unique anomaly that was not observed with any other solvent (10). The electrolytic conductance of aqueous solutions increases with increase in pressure. The effect is more pronounced at lower temperatures. For all other solvents, the electrical conductivity of solutions decreases with increase in pressure. This unusual behavior of water is due to its peculiar associative properties (11).

Thermal expansion causes liquid water to become less dense as the temperature increases. At the same time, the liquid vapor density increases as the pressure rises. For example, the density of water varies from 1.0 g/cm^3 at room temperature to 0.7 g/cm^3 at 306°C. At the critical point, the densities of the two phases become identical and they become a single fluid, called *supercritical fluid*. The water density at this point is only ~0.3 g/cm^3 (Figs. 1.4 and 1.5).

In the region of supercritical point, most properties of supercritical water vary widely. The most prominent of these is the heat capacity at constant pressure, which approaches infinity at the critical point. Even 25°C above T_c, at 80 bar away from P_c the heat capacity of water is an order of magnitude greater than its value at higher or lower pressure.

The dielectric constant of dense, supercritical water can range from 5 to 20 simply on variation of the applied pressure.

As the temperature increases, from ambient to the critical point, the electrolytic conductance of water rises sharply almost independent of pressure. Macroscopically, this is due to the decrease in water viscosity over this range. The primary cause for the fall in viscosity is a disintegration of water clusters.

However, the conductances, which are as much as tenfold greater than at room temperature, begin to drop off once entering the super-

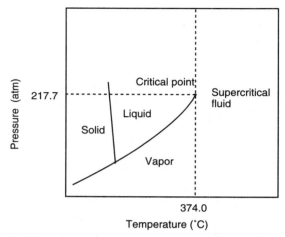

Figure 1.4 Phase diagram of water around the supercritical region.

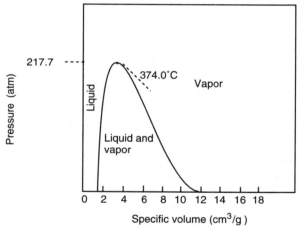

Figure 1.5 Pressure–volume diagram for water around the supercritical region.

critical region. The degree of drop is dependent on the pressure. Low pressures bring about a sharp drop while at high pressures, the decline is much less severe. Dissolved salts associate with themselves and behave as weak electrolytes.

On the other hand, water itself, which is an extremely weak electrolyte at room temperature, dissociates to a greater extent as the temperature rises [Eq. (1.2)]. For example, K_w increases to about 10^{-6} at 1000°C and a density of 1.0 g/mL.

$$2 H_2O \rightleftharpoons H_3O^+ + {}^-OH \qquad (1.2)$$

The increased dissociation of water in conjunction with the increased association of the electrolyte in the supercritical region have a fundamental influence on chemical reactions. The speed of some reactions, such as hydrolysis, increases in supercritical water. For example, there are at least eight species—KCl, KOH, HCl, HOH, K^+, Cl^-, H^+, and OH^-,—for potassium chloride in supercritcal water.

On the other hand, the increase in temperature decreases the intermolecular interaction (hydrogen bonding) between water molecules, which lessens the squeezing-out effect for nonpolar solutes. At the supercritcal state, water exhibits an "antiaqueous property." For example, water at high temperatures exhibits considerable, and sometimes complete, miscibility with nonpolar compounds.

Because oxygen, carbon dioxide, methane, and other alkanes are completely miscible with dense supercritical water, combustion can occur in this fluid phase. Both flameless oxidation and flaming combustion can take place. This leads to an important application in the treatment of organic hazardous wastes. Nonpolar organic wastes such as polychlorinated biphenyls (PCBs) are miscible in all proportions in supercritical water and, in the presence of an oxidizer, react to produce primarily carbon dioxide, water, chloride salts, and other small molecules. The products can be selectively removed from solution by dropping the pressure or by cooling. Oxidation in supercritical water can transform more than 99.9% of hazardous organic materials into environmentally acceptable forms in only a few minutes. A supercritical water reactor is a closed system that has no emissions to the atmosphere, which is different from an incinerator.

Quantum-chemical calculations suggest that supercritical water can provide new reaction pathways by forming structures with the reacting molecule that lower the activation energies for bond breakage and formation (9c). The calculation shows that the more water molecules participate in the reaction, the lower the activation energy of the reaction is. For example, calculation of the following gas-shift reactions [Eq. (1.3)] shows that the activation energy for the first step of the reaction is 61.7 kcal/mol if no additional water molecules participate in the reaction ($n = 0$), whereas the participation of an additional water molecule ($n = 1$) would lower the activation energy to half its original value

(35.6 kcal/mol). A similar decrease in activation energy by the participation of an additional water molecule was found for the second step, and participation of more water molecules in the reaction would further reduce the activation energy.

$$CO + (n+1)\,H_2O \longrightarrow HCOOH + n\,H_2O \longrightarrow CO_2 + H_2 + n\,H_2O \tag{1.3}$$

$$\text{Step 1} \qquad\qquad\qquad \text{Step 2}$$

Water here is effectively acting as a catalyst for the reaction by lowering the energy of activation. These catalytic water molecules are more likely to participate in the reaction under supercritical conditions because their high compressibility promotes the formation of solute–solvent clusters.

REFERENCES

1. Endress, M.; Bischoff, A.; Zinner, E. *Nature* **379**, 701 (1996).
2. Eisenberg, D.; Kauzmann, W. *The Structure and Properties of Water*, Oxford University Press, 1969.
3. Rideout, D. C.; Breslow, R. *J. Am. Chem. Soc.* **102**, 7816 (1980).
4. Horne, R. A., ed. *Water and Aqueous Solutions*, Wiley-Interscience, New York, 1972.
5. Dojlido, J. R.; Best, G. A. *Chemistry of Water and Water Pollution*, Ellis Horwood, 1993.
6. *CRC Handbook of Chemistry and Physics*, 75th ed., CRC Press, Boca Raton, 1994; also Reference 5.
7. Tanford, C. *The Hydrophobic Effect: Formation of Micelles and Biological Membranes*, 2nd ed., Wiley, 1980; Blokzijl, W.; Engberts, J. B. F. N. *Angew. Chem., Int. Ed. Engl.* **32**, 1545 (1993).
8. Dack, M. R. J. *Chem. Soc. Rev.* 211 (1975).
9. (a) Brummer, S. B.; Gancy, A. B. in *Water and Aqueous Solutions: Structure, Thermodynamics, and Transport Processes*, Horne, R. A., ed., Wiley-Interscience, 1972; (b) Gancy, A. B. ibid.; (c) Shaw, R. W.; Brill, T. B.; Clifford, A. A.; Eckert, C. A.; Franck, E. U. *C & EN*, Dec. 23, 1991, p 26,; (d) Reference 2.
10. Cohen, E. *Piezochemie, Kondensierter Systeme, Akademische Verlagsgesellschaft*, Leipzig, 1919.
11. Kavanau, J. L. *Water and Solute-Water Interactions*, Holden-Day, San Francisco, 1964.

CHAPTER 2

PERICYCLIC REACTIONS

When the mind seeks after the essential, it is often confronted with the unsought accidental.

—Plato (427–347 B. C.)*

Pericyclic reactions are those reactions that occur by a concerted process through a cyclic transition state. Such reactions include cycloadditions, sigmatropic rearrangements, and electrocyclizations (1). During such reactions, there is a large decrease in the volume of activation ($\Delta V^{\neq}$) in forming the transition state. For example, the value of $\Delta V^{\neq}$ is between -20 and -45 cm^3/mol for Diels–Alder reactions and -10 cm^3/mol typically for Cope and Claisen rearrangements (2). This large volume contraction can be translated into subtraction of several kilocalories per mole from the free energy of activation under high pressure. Thus, the rates of these reactions are particularly susceptible to a change in pressure. For example, a Diels–Alder cycloaddition reaction under a pressure of 9–10 kbar at room temperature has approximately the same rate as one at about 100°C at atmospheric pressure (3). In addition, whereas an increase in temperature increases the rate of both forward and reverse reactions, the increase in preasure accelerates the forward reaction only.

An analogy can be drawn between pericyclic reactions in water and under high pressure. Water's internal pressure on hydrophobic sub-

*N. Stockhammer, ed., *Plato Dictionary*, Philosophical Library, New York, 1963.

strates acts on the volume of activation of a reaction in the same way as an externally applied pressure does. Thus, the internal pressure of water influences the rates of pericyclic reactions in water in the same direction as do external pressures. The use of salting-out salts will further increase the rate of pericyclic reactions. Recently, Kumar quantified the relationship between internal pressure and the rate of the aqueous Diels–Alder reaction. A linear relationship between the two was observed (4).

On the other hand, many pericyclic reactions are accelerated by Lewis acid catalysts. This acceleration has been attributed to complex formation between the Lewis acid and the polar groups of the reactants, bringing about changes in the energies and orbital coefficients of the frontier orbitals (5). The complex formation also stabilizes the enhanced polarized transition state.

The effect of water molecules on pericyclic reactions can also be compared with the effects of Lewis acids on these reactions. The enhanced polarization of the transition state in these reactions would lead to stronger hydrogen bonds at the polar groups of the reactants, which will result in a substantial stabilization of the transition states in the same way as Lewis acids do. A computer simulation study on the Diels–Alder reaction of cyclopentadiene by Jorgensen indicated that this effect contributes about a factor of 10 to the rates (6).

For intermolecular pericyclic reactions, the aggregation of nonpolar reactants in water (hydrophobic effect) results in a net gain in free energy. This gain in free energy would contribute to the increase in the rate of reactions. Breslow has studied the influence of the hydrophobic effect on the aqueous Diels–Alder reactions in detail (7, 8). Schneider has reported a quantitative correlation between solvophobicity and rate enhancement of aqueous Diels–Alder reactions (9, 10). Enforced hydrophobic interactions between diene and dienophile have also been suggested by Engberts to account for the acceleration in water (11). A pseudothermodynamic analysis of the rate acceleration in water relative to 1-propanol and 1-propanol/water mixtures indicates that (1) hydrogen-bond stabilization of the polarized activated complex and (2) the decrease of the hydrophobic surface area of the reactants during the activation process are the two main causes of the rate enhancement in water (12). Studies of Diels–Alder reactions under high pressure by Jenner revealed that water can alter kinetics and chemo- and enantioselectivity through polarity and hydrophobic effects (13). The weight of factors 1 and 2 (above), however, depends on the use of specific reaction partners. By studying the reaction in dilute aqueous ethanol, Smith

(14) and Griesbeck (15) pointed out that the concentration of the reaction substrates is important for rate enhancement. The rate shows a maximum at 0.5 M under the aqueous ethanol condition.

The influence of the hydrophobic effect on aqueous pericyclic reactions can be compared with the effect of catalytic antibodies. Antibodies have been found to catalyze Diels–Alder reactions (16), hetero-Diels–Alder reactions (17), and Claisen (18) and oxy-Cope (19) rearrangements. It has been suggested that antibodies catalyze these reactions by acting as an entropy trap, primarily through binding and orienting the substrates in the cyclic conformations.

2.1 DIELS–ALDER REACTIONS

The Diels–Alder reaction (20) is one of the most important methods used to form cyclic structures. Diels–Alder reactions in aqueous media were first carried out in the 1930s (21). No special attention was paid to this fact until recently. In 1980, Breslow (22) made the first dramatic observation that the reaction of cyclopentadiene with butenone in water was more than 700 times faster than the same reaction in isooctane. The reaction rate in methanol is comparable to that in a hydrocarbon solvent. Such an unusual acceleration of the Diels–Alder reaction by water was attributed to the "hydrophobic effect" (23), in which the hydrophobic interactions brought together the two nonpolar groups in the transition state (Fig. 2.1).

The use of β-cyclodextrin, which simultaneously forms an inclusion complex with the diene and dienophile, and the use of 4.86 M LiCl aqueous solution as solvent, which salts out nonpolar materials dissolved in water (24), has been shown to further enhance the rate of aqueous Diels–Alder reactions (Table 2.1). On the other hand, the use of α-cyclodextrin decreases the rate of the reaction. This inhibition was explained by the fact that the relatively smaller cavity can only accommodate the binding of cyclopentadiene, leaving no room for the

Figure 2.1 Transition state of Diels–Alder reaction between cyclopentadiene and butenone.

TABLE 2.1 Rate Constants for Selected Diels–Alder Reactions

Solvent	Substrates	$k_2 \times 10^5$ (M^{-1} s^{-1})
Isooctane	A^a	5.94 ± 0.3
MeOH	A	75.5
H_2O	A	$4{,}400 \pm 70$
H_2O/LiCl (4.86 M)	A	10,800
H_2O/$C(NH_2)_3^+Cl^-$ (4.86 M)	A	4,300
H_2O/β-CD (10 mM)	A	10,900
H_2O/α-CD (10 mM)	A	2,610
H_2O	B^b	$22{,}600 \pm 700$
CH_3CN	B	107 ± 8
H_2O/β-CD (10 mM)	B	13,800

^{a}Cyclopentadiene + butenone.
^{b}Anthracene-9-carbinol + N-ethylmaleimide.

dienophile. Similar results were observed between the reaction of cyclopentadiene and acrylonitrile.

The reaction between hydroxymethylanthracene and N-ethylmaleimide was also studied (Table 2.1). In water at 45°C, its second-order rate constant in water was over 200 times larger than in acetonitrile [Eq. (2.1)]. In this case, the β-cyclodextrin became an inhibitor, rather than an activator, due to the even larger transition state, which cannot fit into its cavity. A slight deactivation was also observed with a salting-in salt solution, such as quanidinium chloride aqueous solution.

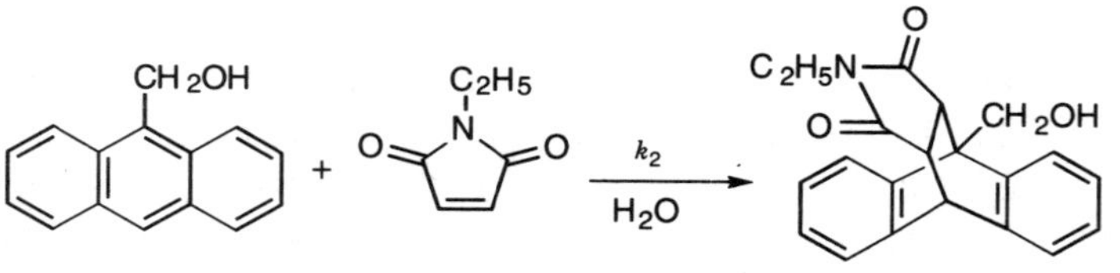

$$\text{(2.1)}$$

Water as a reaction solvent also strikingly affected the stereoselectivity of some Diels–Alder reactions (25). At low concentrations, where both components were completely dissolved, the reaction of cyclopentadiene with butenone gave a 21.4 ratio of endo/exo products when they were stirred at 0.15 M concentration in water, compared to only a 3.85 ratio in excess cyclopentadiene and an 8.5 ratio in ethanol as the solvent [Eq. (2.2)]. An aqueous detergent solution had no effect on the product ratio. The stereochemical changes could be explained by the need to minimize the transition-state surface area in water solution, thus favoring the more compact endo stereochemistry. The results are also consistent with the effect of polar media on the ratio (26).

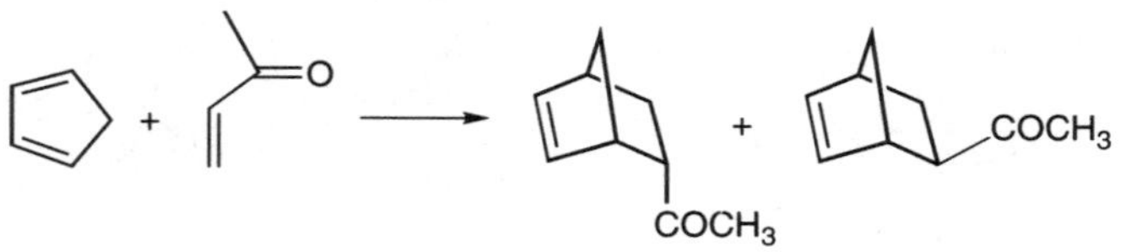

$$(2.2)$$

Medium	Endo/exo ratio (25°C)
Cyclopentadiene	3.85
Ethanol	8.5
Water	21.4
Water + surfactant	19.5

A similar water-induced selectivity was observed in the reaction of cyclopentadiene with dimethyl maleate or methyl methacrylate. Even though the bulk of both diene and dienophile is poorly soluble and present as a separate phase in the maleate case, the influence of water on the selectivity is still phenomenal.

The catalytic behavior exhibited by β-cyclodextrin was also observed by Sternbach in the intramolecular Diels–Alder reaction of a furan-ene in water [Eq. (2.3)] (27). In water alone, the cyclization proceeded in 20% yield with an epimeric selectivity of $1:2$ (**A**:**B**) at 89°C after 6 h. The same reaction gave 91% of the cyclized product in the presence of one equivalent of β-cyclodextrin. In this case, the epimeric selectivity was also changed to $1:1.5$ (**A**:**B**). However, no significant change of reactivity was observed with either α-cyclodextrin or the nonionic detergent, Brij-35, present.

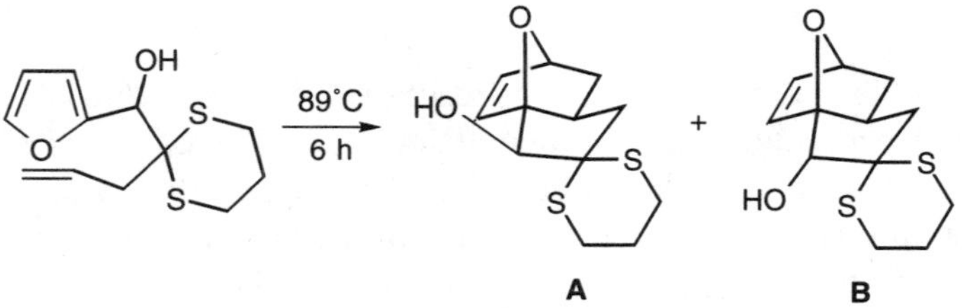

$$(2.3)$$

Solvent (equiv)	A/B	Yield (%)
Water	1:2	20
Water + α-CD (1.0)	—	25
Water + β-CD (1.0)	1:1.5	91
Water + Brij-35 (1.0)	—	22

Two complexation models have been suggested for the enhanced intramolecular Diels–Alder reaction (**A** and **B**). Either case, however,

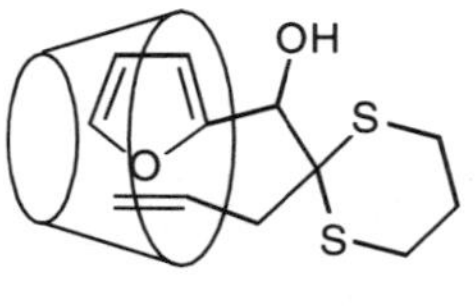

would result in the closing up of the two reactive ends, leading to an increased rate in the cyclization. A similar enhancement of reactivity by β-cyclodextrin was observed in the cyclization of the amine derivative [Eq. (2.4)]. Roskamp reported a similar intramolecular Diels–Alder reaction accelerated by silica gel saturated with water [Eq. (2.5)] (28). The reaction led to the ready construction of the bicyclo [6,2,1] ring systems. Intramolecular Diels–Alder reaction has also been investigated by Keay (29). A Diels–Alder reaction of 2,5-dimethylpyrrole derivatives with dimethyl acetylenedicarboxylate in water generated the corresponding cyclization products (30).

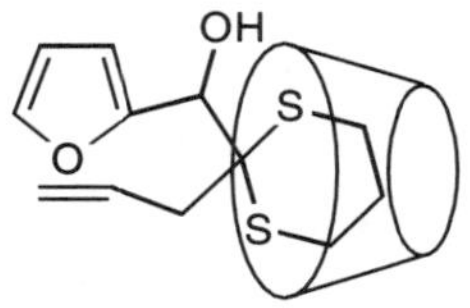

Holt studied the Diels–Alder reaction in a mixture of water, 2-propanol, and toluene as microemulsions (31). The endo/exo ratio between the reaction of cyclopentadiene and methyl methacrylate was enhanced with increasing amount of water in the presence of a surfactant.

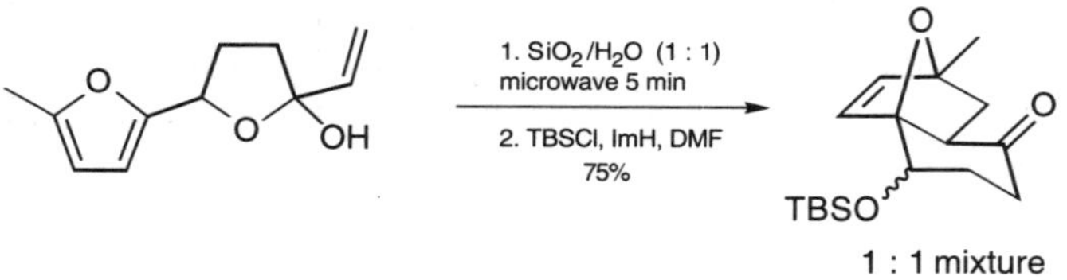

In spite of many examples of acceleration of Diels–Alder reactions by the use of aqueous media, Elguero et al. reported (32) that the Diels–Alder reaction between cyclopentadiene and methyl (and benzyl) -2-acetamideacrylates proceeded better in toluene than in water in both yield and exo/endo selectivity. Ultrasonic irradiation did not improve the yield.

Grieco has made an important contribution in studies of aqueous Diels–Alder reactions toward the syntheses of a variety of complex natural products. When the Diels–Alder reaction in Scheme 2.1 was carried out in water, a higher reaction rate and reversal of the selectivity were observed, than in the same reaction in a hydrocarbon solvent (33). It should be noted that, for the aqueous reaction, the sodium salt of the diene was used. Water-soluble cosolvents caused a rapid reduction in rate. The best result was obtained when the reaction was conducted with a fivefold excess of the sodium salt of diene carboxylate (see Scheme 2.1).

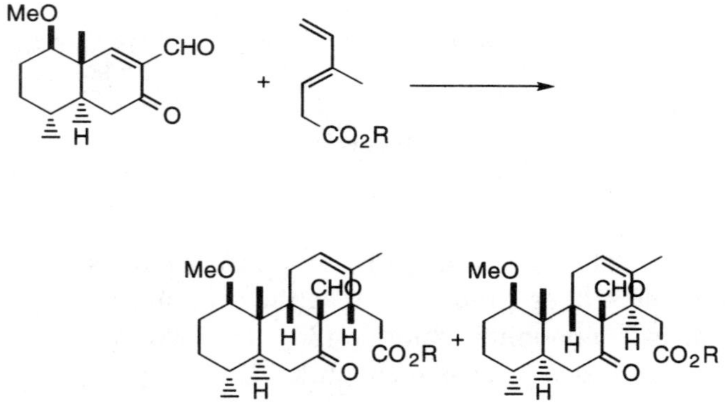

Ratio **A/B** = 3 : 1 (in water)   R = Na
              1 : 0.85 (in benzene)   R = Et

Scheme 2.1

Similar results were obtained with the compounds shown in Scheme 2.2. In this case, the reaction in water at room temperature is faster than the same reaction in toluene at 100°C. Enhanced selectivity (3:1) was also observed in water in comparison to the reaction in toluene (1:1.1). Such an enhancement can also be explained by the hydrophobic effect.

The Diels–Alder reaction of the dienecarboxylate in water was also investigated with a variety of other dienophiles (Table 2.2) (34). Substantial enhancement of the reaction rate was also observed. Sensitive dienol ether functionality in the diene carboxylate was shown to be compatible with the conditions of the aqueous Diels–Alder reaction

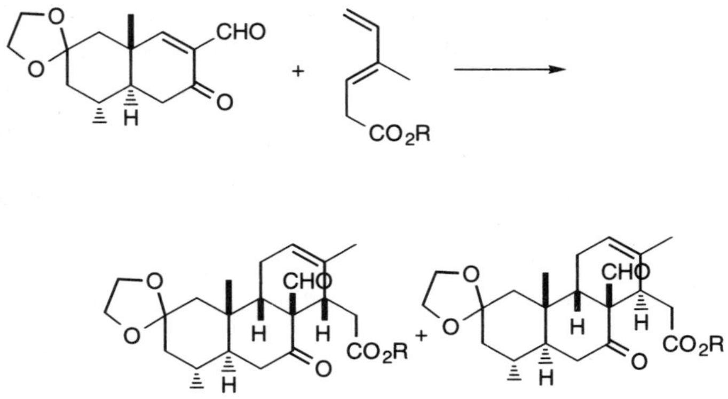

Solvent	Temp. (°C)/Time/Yield(%)
Water	RT/4.5 h/90
Toluene	100/36 h/97

Scheme 2.2

[Eq. (2.6)] (35). The dienes in the Diels–Alder reactions can also bear other water-solubilizing groups, such as the sodium salt of phosphoric acid, and dienyl ammonium chloride (36). The hydrophilic acid functionality can also locate at the dienophile (37).

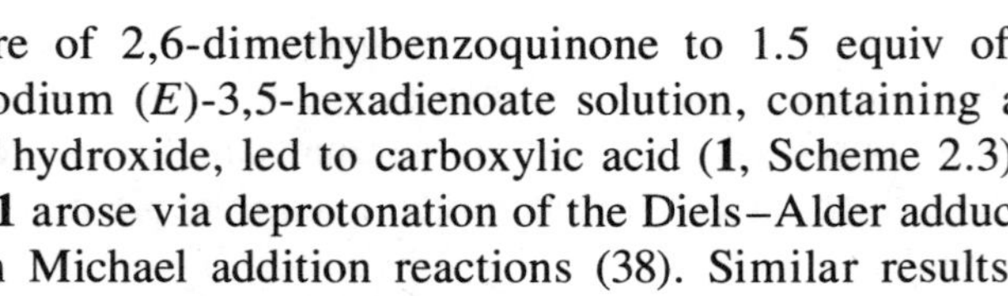

$$(2.6)$$

Exposure of 2,6-dimethylbenzoquinone to 1.5 equiv of a 1.0 M aqueous sodium (E)-3,5-hexadienoate solution, containing a catalytic amount of hydroxide, led to carboxylic acid (**1**, Scheme 2.3). The formation of **1** arose via deprotonation of the Diels–Alder adduct followed by tandem Michael addition reactions (38). Similar results were obtained with sodium (E)-4,6-heptadienoate.

Cycloaddition of sodium (E)-3,5-hexadienoate with an α-substituted acrolein in water followed by direct reduction of the intermediate Diels–Alder adduct gave **2** in excellent yield in one pot (39). This com-

TABLE 2.2 Diels–Alder Reaction with Sodium (*E*)-4,6-Heptadienoate

Substrate	Product	Yield (%)
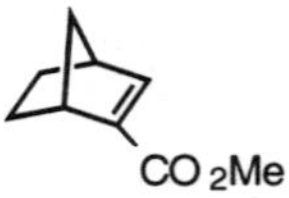	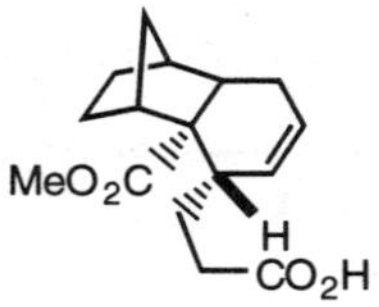	94
CH$_2$=CHCHO	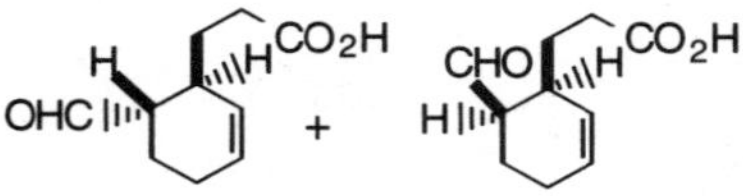	70
CH$_2$=C(Me)CHO	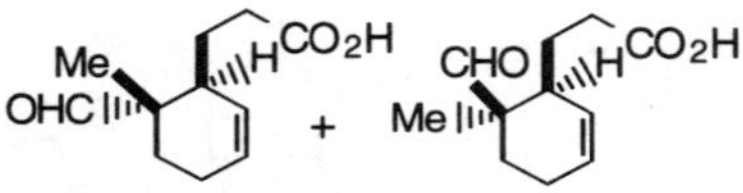	67
CH$_2$=CH(NO$_2$)Me	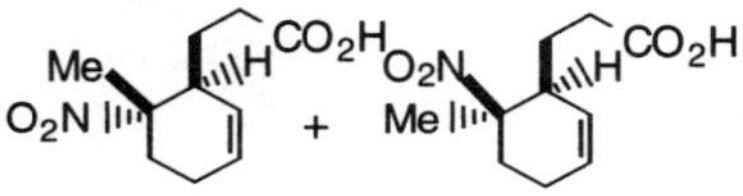	51
MeO$_2$C—≡—CO$_2$Me	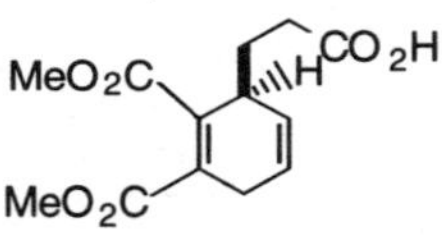	60
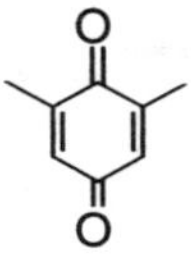	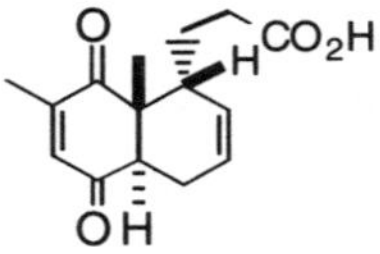	86
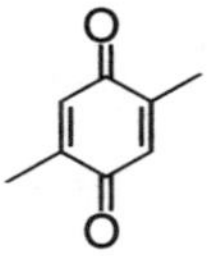	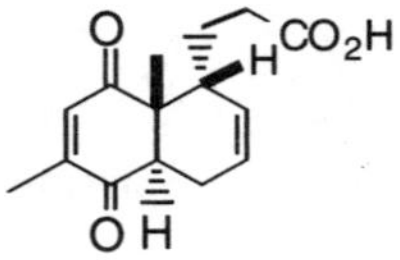	78
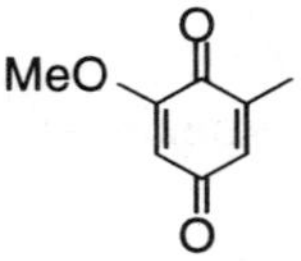	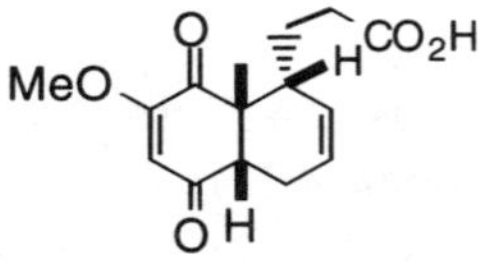	79
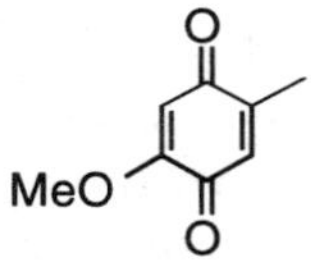	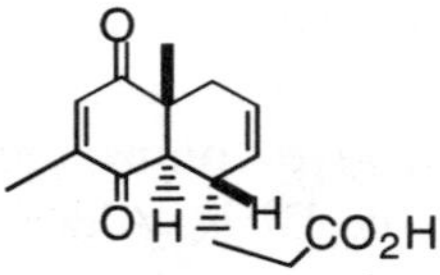	53

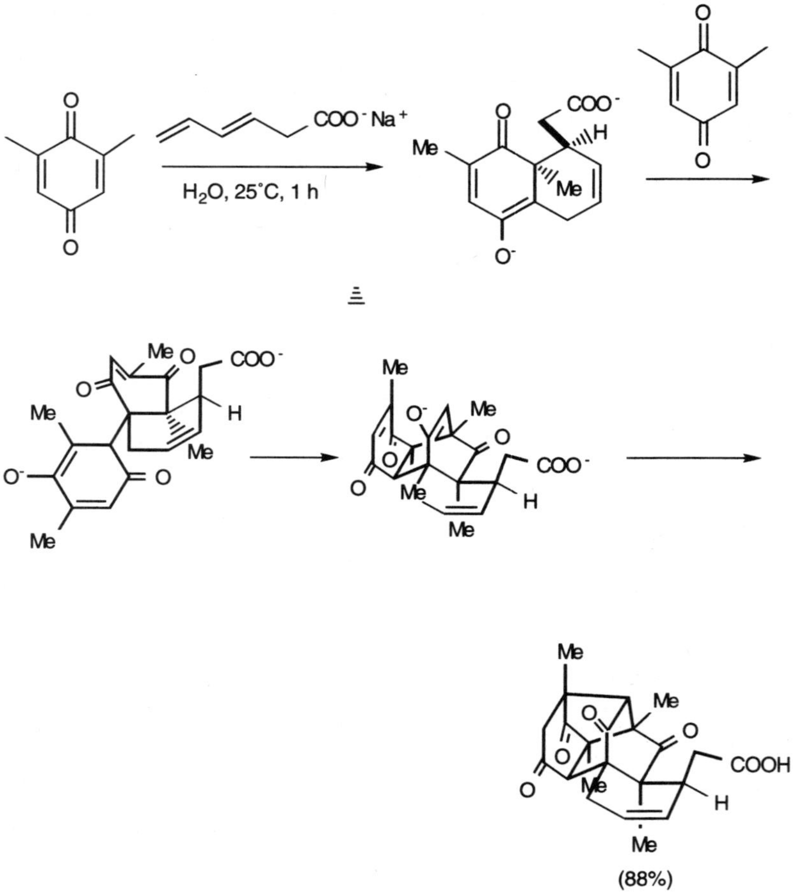

Scheme 2.3

pound constituted the basic **AB** ring system of the sesquiterpene lactone, vernolepin (**3**) (Scheme 2.4).

Similar reactions were applied to the syntheses of *dl-epi*-pyroangolensolide, *dl*-pyroangoensolide (**4**) (40), and the Inhoffen–Lythgoe diol (**5**) (41). Sodium (*E*)-4, 6, 7-octatrienoate reacted smoothly with a variety of dienophiles to give conjugated diene products (42). An intramolecular version of the Diels–Alder reaction with a dienecarboxylate was used by Williams et al. in the synthetic study of the antibiotic ilicicolin H (**6**) (43).

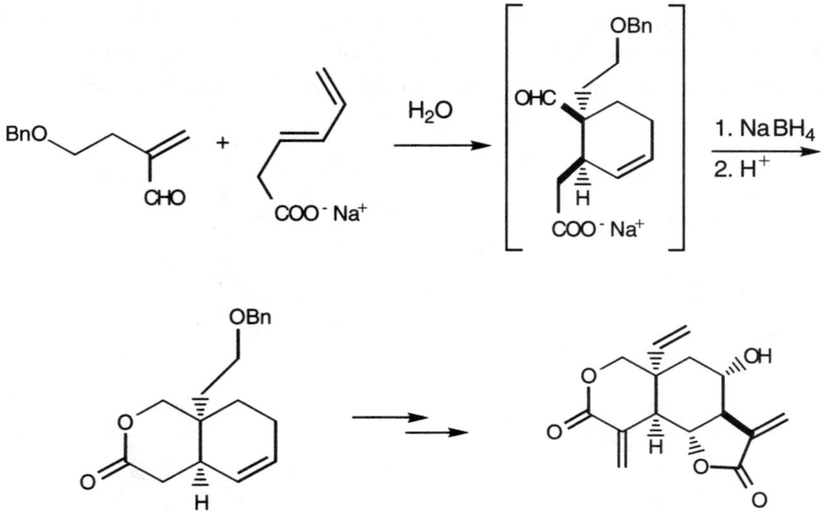

Scheme 2.4

91% overall
2

Vernolepin
3

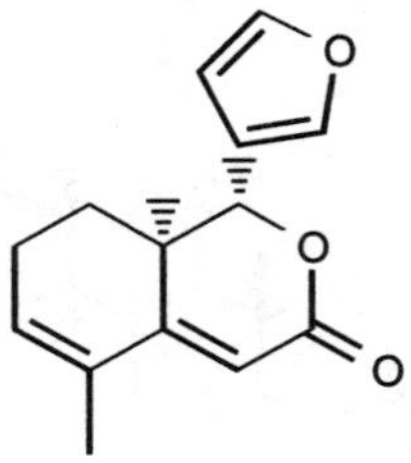

Pyroangolensolide
4

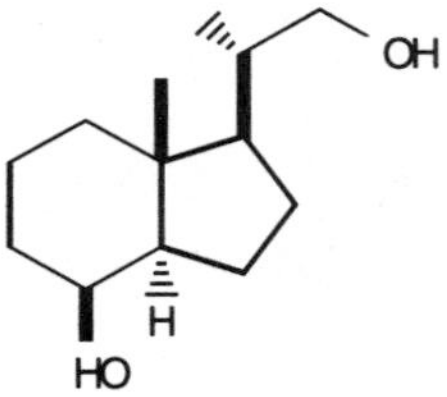

Inhoffen–Lythgoe diol
5

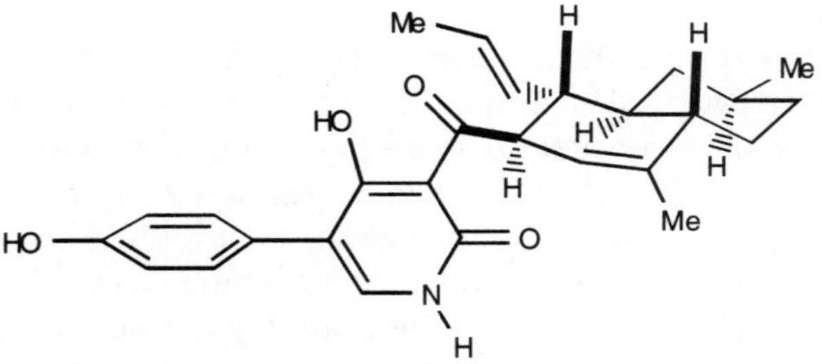

6   Ilicicolin H

De Clercq has shown that aqueous Diels–Alder reactions can be used as key steps in the syntheses of (±)-11-ketotestosterone (**7**) (Scheme 2.5) (44) and (±)-gibberellin A₅ (**8**) (Scheme 2.6) (45).

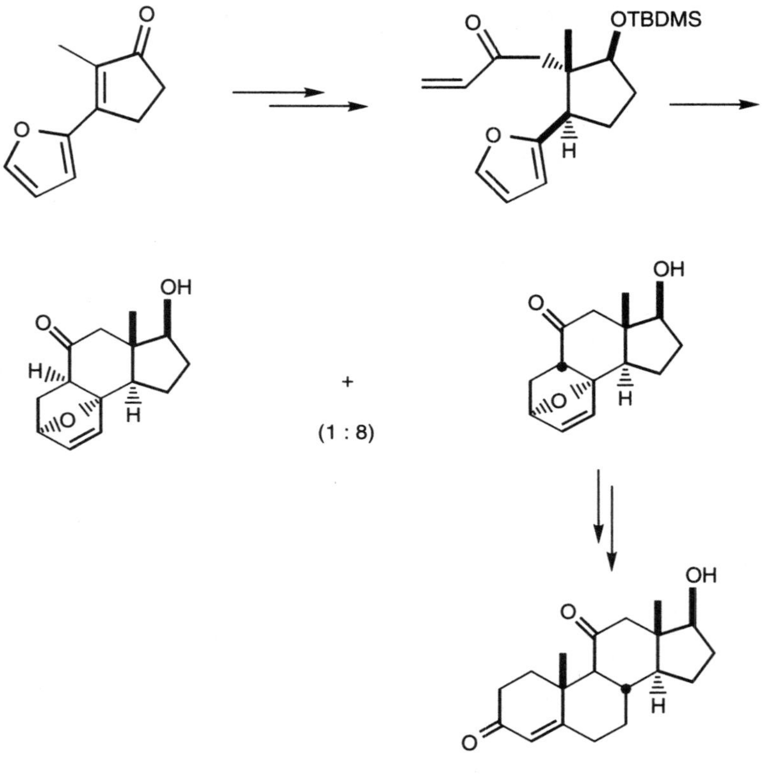

(1 : 8)

7   Adrenosterone

Scheme 2.5

A diene bearing a chiral water-soluble glycohydrophilic moiety was studied extensively by Lubineau (46). The use of water-soluble glycoorganic compounds in water achieved higher reagent concentration. The reaction also resulted in rate enhancement and asymmetric induction. Even though the diastereoselectivity was modest (20% d.e.), separation of the diastereomers led to chiral adducts in pure enantiomeric form after cleavage of the sugar moiety by acidic hydrolysis, or by using glycosidase in neutral conditions at room temperature (Scheme 2.7). A

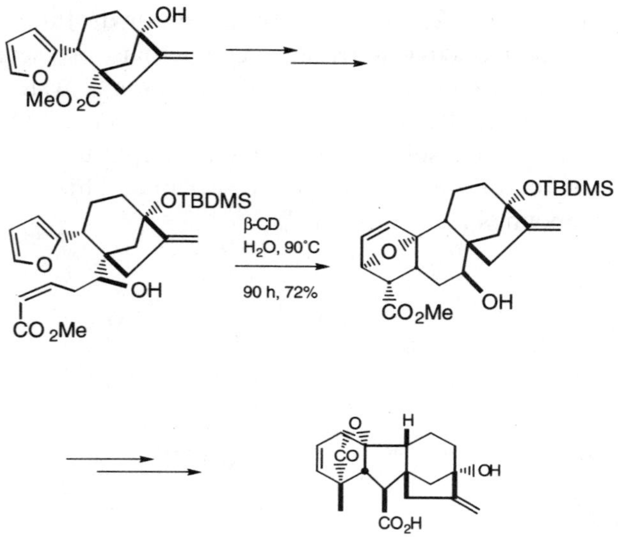

Scheme 2.6

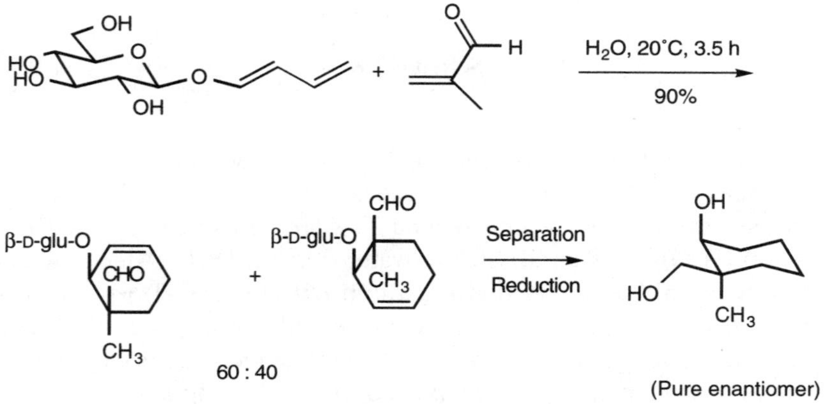

Scheme 2.7

variety of substrates bearing other glyco derivatives was also studied. The reaction of dienes bearing a *N*-dienyl lactam moiety with activated olefins was examined by Smith (47).

In the synthesis of 2,2,5-trisubstituted tetrahydrofurans, a novel class of orally active azole antifungal compounds, Saksena (48) reported that the key step of the Diels–Alder reaction in water led to the desired substrate virtually in quantitative yield (Scheme 2.8), while the same reaction in organic solvent resulted in a complicated mixture with only <10% of the desired product being isolated. This success made the target compounds readily accessible.

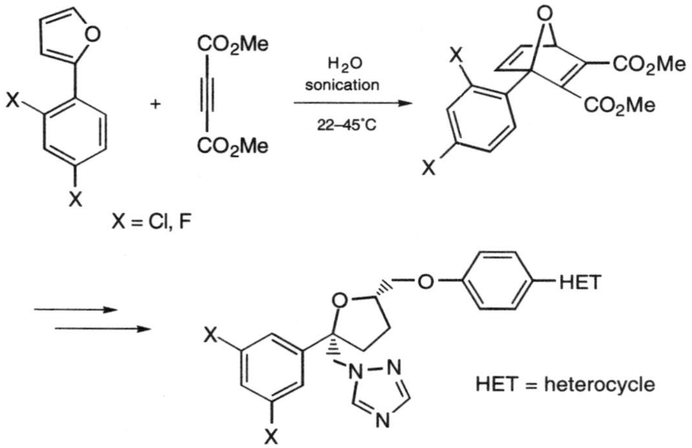

Scheme 2.8

An important aspect of the Diels–Alder reaction is the use of Lewis acids for the activation of the substrates. While most Lewis acids are decomposed or deactivated in water, Bosnich reported that $[Ti(Cp^*)_2(H_2O)_2]^{2+}$ is an air-stable, water-tolerant Diels–Alder catalyst (49). Kobayashi has found that scandium triflate $[Sc(OTf)_3]$ (50), and lanthanide triflates, $[Ln(OTf)_3]$, are stable and can be used as a Lewis catalyst under aqueous conditions. Other simple metal salts can be used as well. For example, Engberts reported (51) that the following cyclization reaction [Eq. (2.7)] in an aqueous solution containing 0.010 M

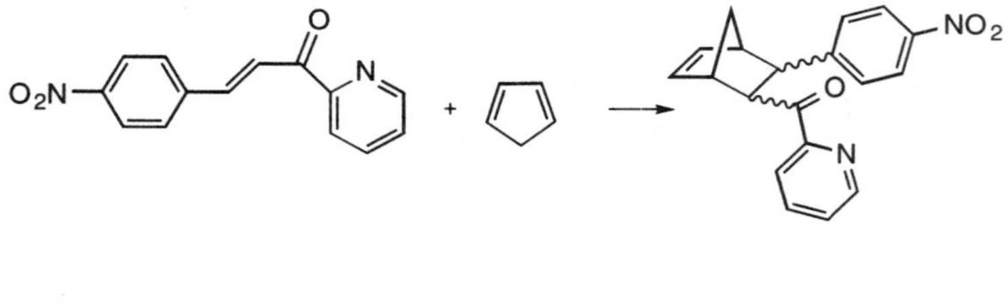

(2.7)

$Cu(NO_3)_2$ is 250,000 times faster than that in acetonitrile and ~1,000 times faster than that in water alone. Other salts, such as Co^{2+}, Ni^{2+}, Zn^{2+}, also catalyze the reaction, but are not as reactive as Cu^{2+}. The reaction is also catalyzed by bovine serum albumin (52).

The Diels–Alder reaction between oxazolone and cyclopentadiene was investigated by Cativiela in water (Scheme 2.9) (53). The cycloadducts are readily converted into aminonorbornane carboxylic acids.

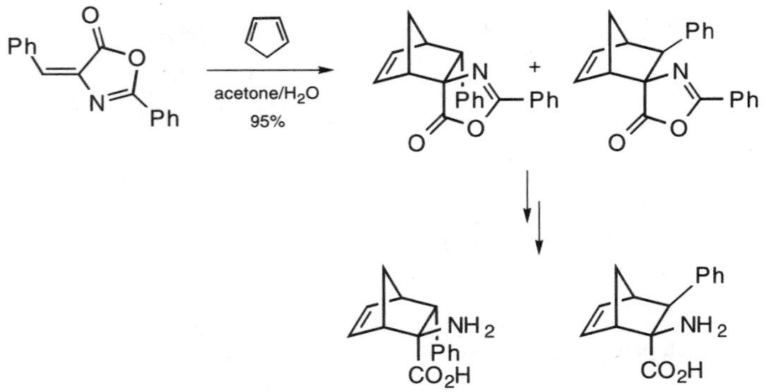

Scheme 2.9

2.1.1 Hetero-Diels–Alder Reactions

For the synthesis of heterocyclic compounds, hetero-Diels–Alder reactions with nitrogen- or oxygen-containing dienophiles are particularly useful (54). In 1985, Grieco reported the first example of hetero-Diels–Alder reactions with nitrogen-containing dienophiles in aqueous media. Simple iminium salts, generated *in situ* under Mannich-like conditions, reacted with dienes in water to give aza-Diels–Alder reaction products with the potential for alkaloid synthesis (Table 2.3) [Eq. (2.8)] (55). The use of alcoholic solvents led to a decrease in the reaction rate. The use of THF as a cosolvent did not affect the rate of the reaction.

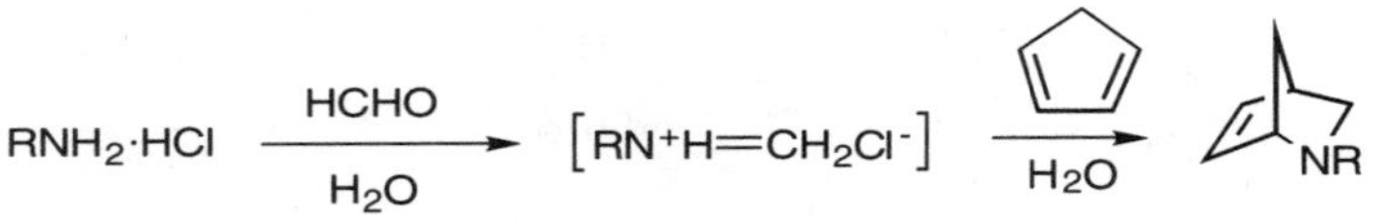

The intramolecular aza-Diels–Alder reaction (56) occurs similarly in aqueous media. The reactions lead to the formation of fused-ring

TABLE 2.3 Aza Diels–Alder Reactions in Aqueous Solution

Diene	Amine + Carbonyl Compound	Product	Yield (%)
	$BnNH_2 \cdot HCl$ + HCHO		41
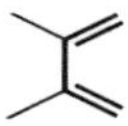	$BnNH_2 \cdot HCl$ + HCHO	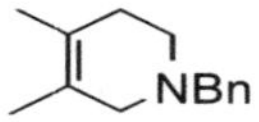	69
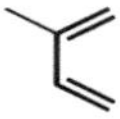	$BnNH_2 \cdot HCl$ + HCHO	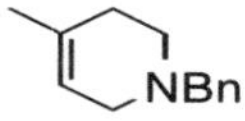	59
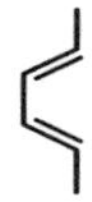	$BnNH_2 \cdot HCl$ + HCHO	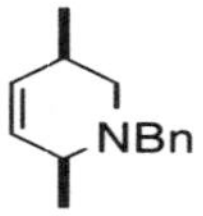	62
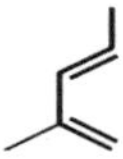	$BnNH_2 \cdot HCl$ + HCHO	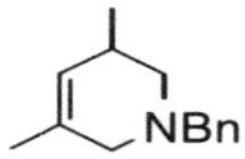	49
	$MeNH_2 \cdot HCl$ + HCHO		82
	$NH_4 \cdot HCl$ + HCHO	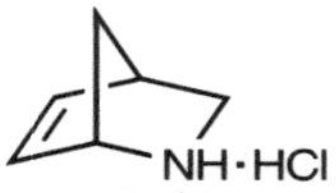	44
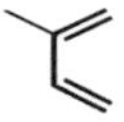	$NH_4 \cdot HCl$ + HCHO	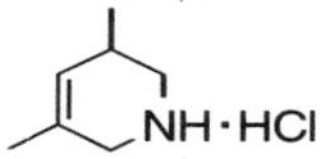	40
	$BnNH_2 \cdot HCl$ + MeCHO		47

systems with bridgehead nitrogen, a structural characteristic of many alkaloids:

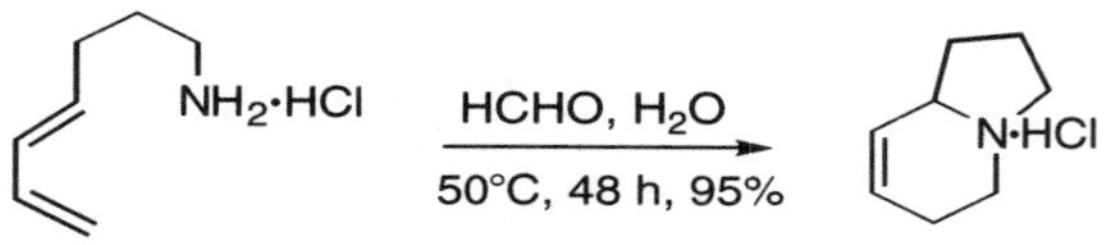

$$(2.9)$$

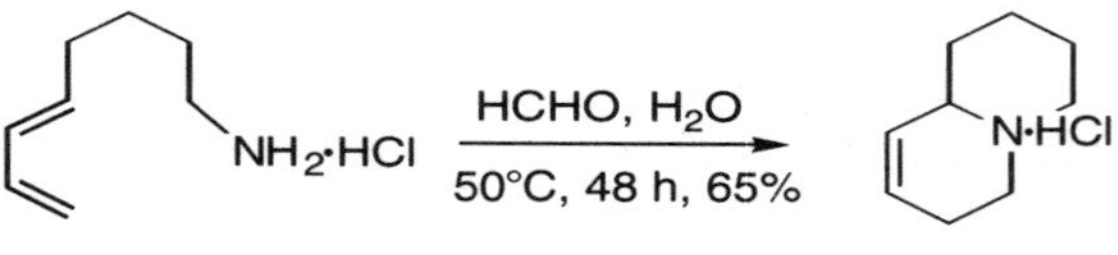

$$(2.10)$$

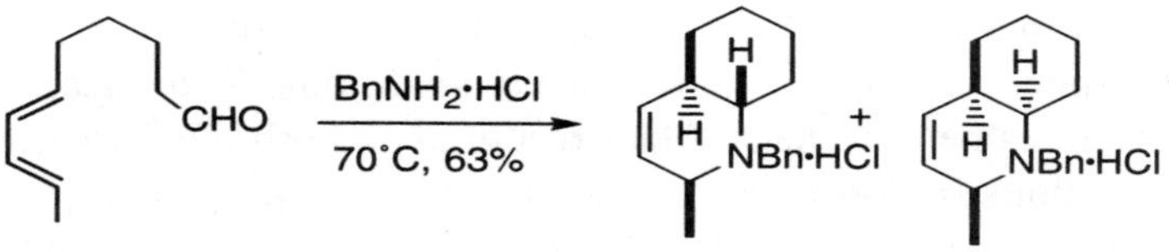

$$(2.11)$$

C-Acyl iminium ions also reacted similarly with cyclopentadienes (Table 2.4) (57).

TABLE 2.4 Aza Diels–Alder Reactions of Cyclopentadiene with C-Acyl Iminium Ions

Substrates	Amine	Product	Yield (%)
C_6H_5COCHO	$CH_3NH_2 \cdot HCl$	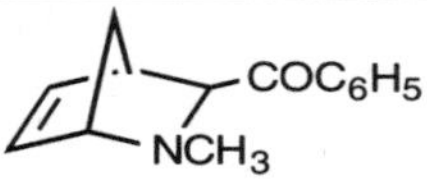	82
CH_3COCHO	NH_4Cl	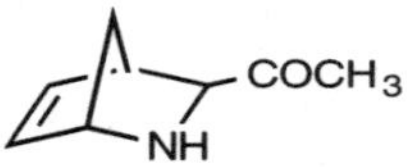	84
CH_3COCHO	$CH_3NH_2 \cdot HCl$	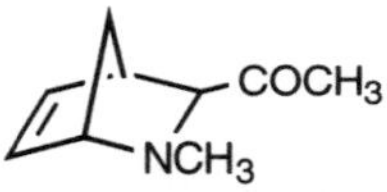	67
CH_3COCHO	$BnNH_2 \cdot HCl$	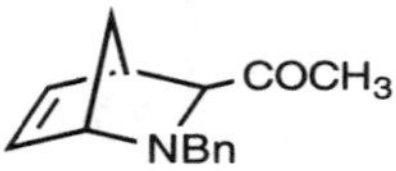	65
$HOOCCHO$	CH_3NH_2	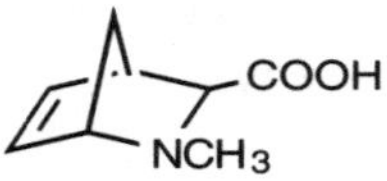	86

Retro aza-Diels–Alder reactions also occurred readily in water (58). In the following example, 2-azanorbornenes undergo acid-catalyzed retro-Diels–Alder cleavage in water [Eq. (2.12)]. The produced iminium

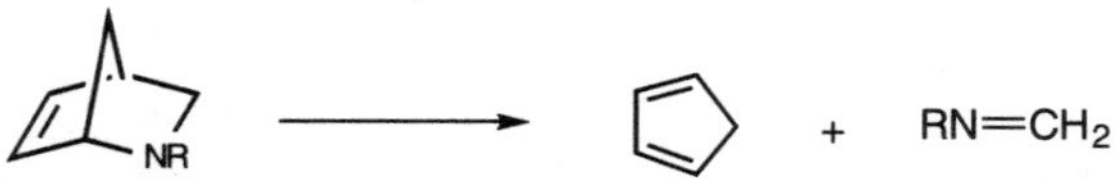

$$(2.12)$$

derivative can react with a trapping reagent, such as N-methylmale-imide, or be reduced to give primary amines [Eq. (2.13)]. No reaction

$$(2.13)$$

was observed in a variety of organic solvents, such as benzene, THF, or acetonitrile under similar or more drastic conditions. This means that water accelerates hetero-Diels–Alder reactions in both the forward and reverse directions by lowering the energy of the transition state. This reaction provided a novel method for the *N*-methylation of dipeptides and amino acid derivatives (Table 2.5) (59).

TABLE 2.5 *N*-Methylation of Peptides and Amino Acid Derivatives

RNH$_2$·HCl	RNMeH	Overall Yield (%)
H-Leu-OMe·HCl	HMe-Leu-OMe·TFA	86
H-Phe-OMe·HCl	Me-Phe-OMe·TFA	77
H-Val-OMe·HCl	Me-Val-OMe·TFA	64
H-Tyr-OMe·HCl	Me-Tyr-OMe	65
L-Phenylglycine methyl ester·HCl	*N*-Methyl-phenylglycine methyl ester	83
H-Lys(Z)-OMe·HCl	Me-Lys(Z)-OMe	81
H-Ser-OMe·HCl	Me-Ser-OMe·TFA	74
H-Leu-Phe-OMe·HCl	Me-Leu-Phe-OMe	68
H-Ala-Ala-Ala-OMe-HOAc	Me-Ala-Ala-Ala-OMe-HOAc	54

Waldmann used (*R*)- and (*S*)-amino acid methyl esters and chiral amines as chiral auxiliaries in analogous aza-Diels–Alder reactions with cyclodienes (Scheme 2.10) (60). The diastereoselectivity of these reactions ranged from moderate to excellent (Table 2.6). Open-chain dienes reacted similarly.

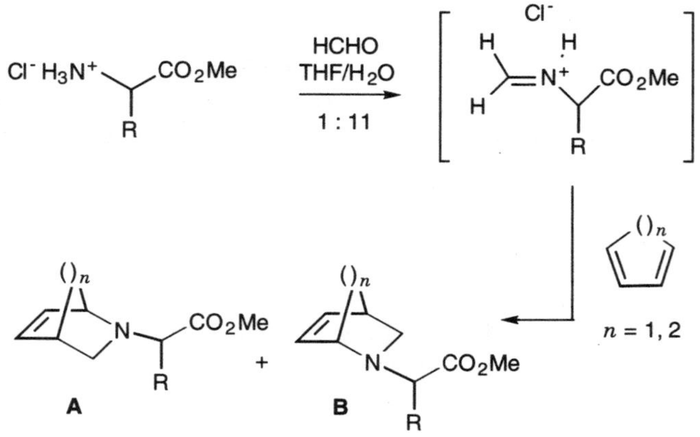

Scheme 2.10

TABLE 2.6 Aza-Diels–Alder Reaction with Amino Acid Derivatives

Aminoacid	Diene	A : B	Yield (%)
(S)-Ile	Cyclopentadiene	93 : 7	57
(S)-Val	Cyclopentadiene	86 : 14	74
(S)-Ser	Cyclopentadiene	87 : 13	63
(S)-Phg	Cyclopentadiene	78 : 22	97
(R)-Val	Cyclopentadiene	17 : 83	69
(R)-Ser	Cyclopentadiene	12 : 88	66
(R)-Phg	Cyclopentadiene	20 : 80	90
(S)-Ile	Cyclohexadiene	80 : 20	35
(R)-Ser	Cyclohexadiene	27 : 63	40

Recently, the aza-Diels–Alder reaction was used by Waldmann in the asymmetric synthesis of highly functionalized tetracyclic indole derivatives [Eq. (2.14)], which is useful for the synthesis of yohimbine- and reserpine-type alkaloids (61). As in the case of Diels–Alder reactions, aqueous aza-Diels–Alder reactions are also catalyzed by lanthanide triflates (62).

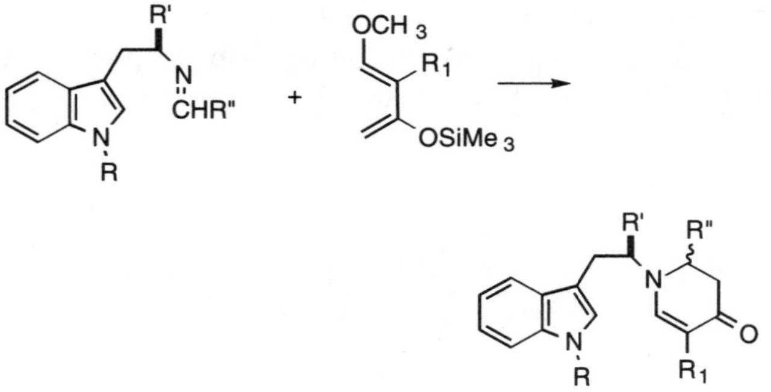

$$(2.14)$$

R = H, CHO
R' = CO$_2$Me, H
R" = phenyl, 4-Cl-phenyl, 4-NO$_2$-phenyl, heptyl, 2-propyl, H.

R$_1$ = H, ethyl

For hetero-Diels–Alder reactions with an oxygen-containing dienophile, cyclopentadiene, or cyclohexadiene reacted with an aqueous solution of glyoxylic acid to give α-hydroxyl-γ-lactones arising from the rearrangement of the cycloadducts (Scheme 2.11). The reaction was independently studied by Lubineau (63) and Grieco (64). Water as a solvent allowed direct use of the inexpensive aqueous solution of glyoxylic

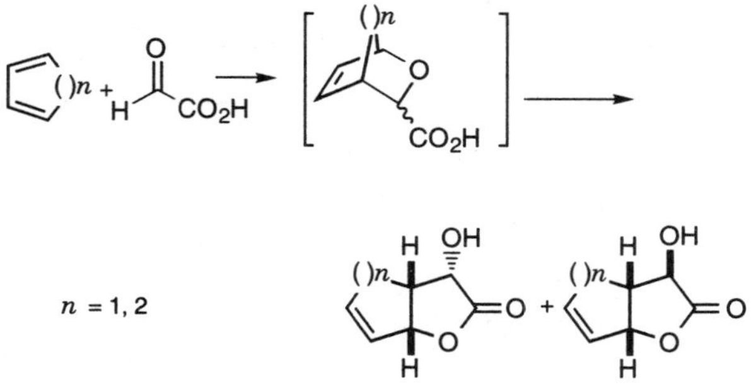

$$n = 1, 2$$

	Yield %	A : B
n = 1	83	73:27
n = 2	85	60:40

Scheme 2.11

acid for cycloaddition, and it also enhanced the rate of the hetero-Diels–Alder reaction relative to the dimerization of cyclopentadiene. The reaction is much faster at low pH, which implies that the reaction is acid-catalyzed.

The 5,5-fused system generated has been used in the total synthesis of several bioactive compounds, including sesbanimides A (**9a**) and B (**9b**), carbovir (**10**) (65), and the hydroxylactone moiety (**11**) of mevinic acids (66).

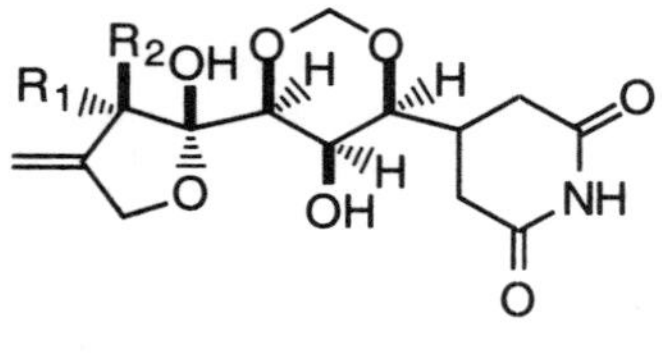

9a $R_1 = H$, $R_2 = Me$ sesbanimide A
9b $R_1 = Me$, $R_2 = H$ sesbanimide B

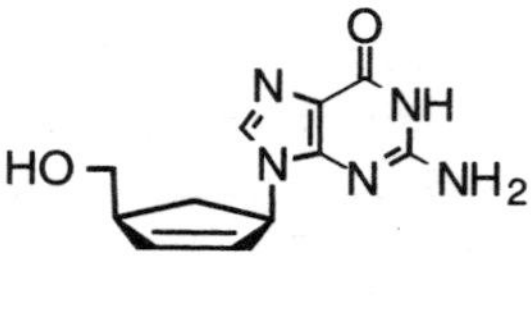

10   carbovir 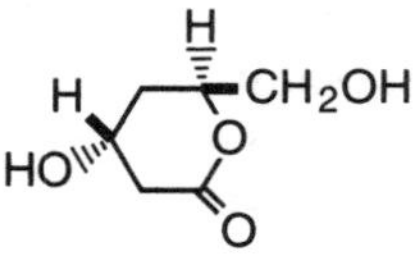   **11**

TABLE 2.7 Oxo-Diels–Alder Reaction with Acyclic Dienes

Diene	Carbonyl Compounds	*syn : anti*	Product	[Yield (%)]
	H—C(=O)—CO_2H	64:36		(57)
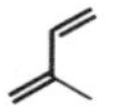	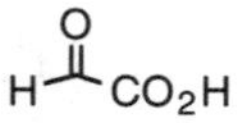 H—C(=O)—CO_2H	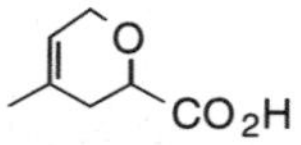		(28)
	H—C(=O)—C(=O)CH₃	53:47		(96)
	H—C(=O)—CHO	50:50		(36)
	CH₃—C(=O)—CO_2H	1:2		(74)

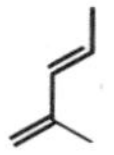

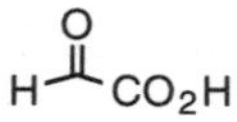

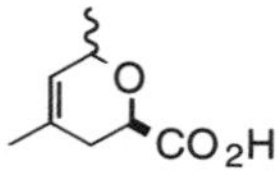

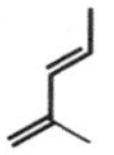

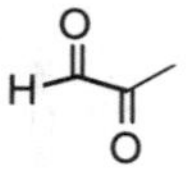

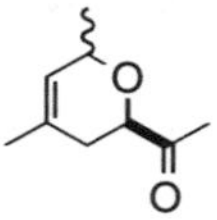

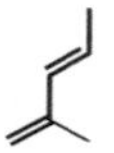

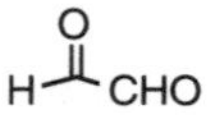

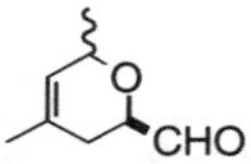

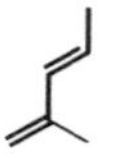

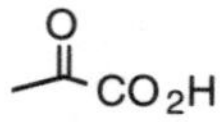

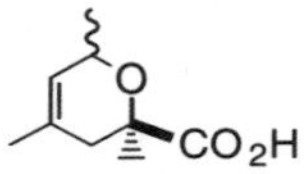

The reaction is not limited to cyclodienes. Acyclic dienes react to give dihydropyran derivatives (Table 2.7). An excellent application of oxo-Diels–Alder reaction is reported by Lubineau et al. in the synthesis of the sialic acids, 3-deoxy-D-manno-2-octulosonic acid (KDO) (Scheme 2.12), 2-deoxy-KDO, and the thioglycoside of KDO (67).

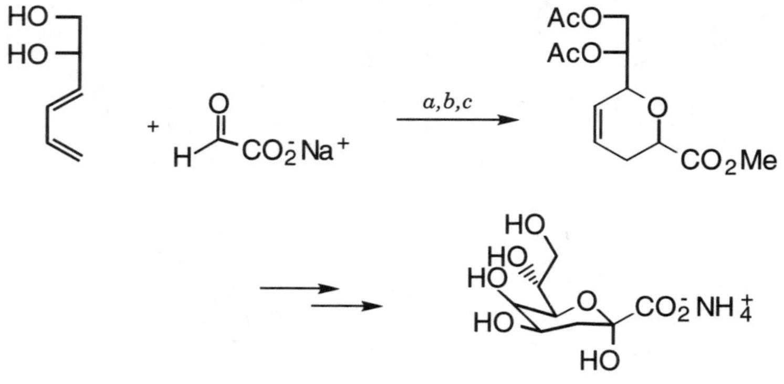

Scheme 2.12 (a) $HCOCO_2^-Na^+$, H_2O; (b) MeOH, H^+; (c) Ac_2O, pyridine.

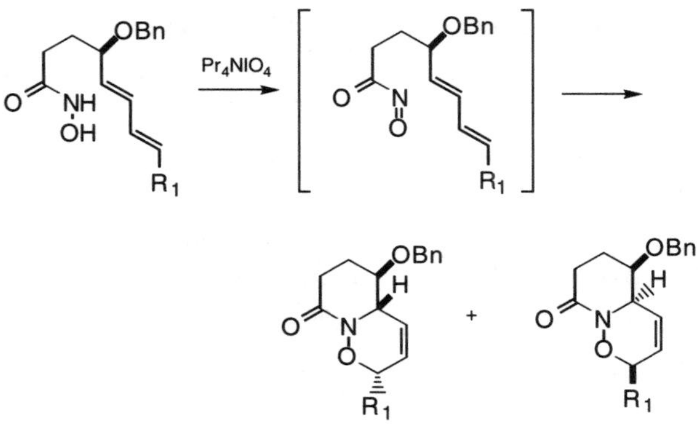

trans *cis*

	Solvent	*Trans/cis*
R_1 = H	$CHCl_3$	1.3 : 1
	H_2O	4.0 : 1
R_1 = Et	$CHCl_3$	1.4 : 1
	H_2O/MeOH	4.5 : 1

Scheme 2.13

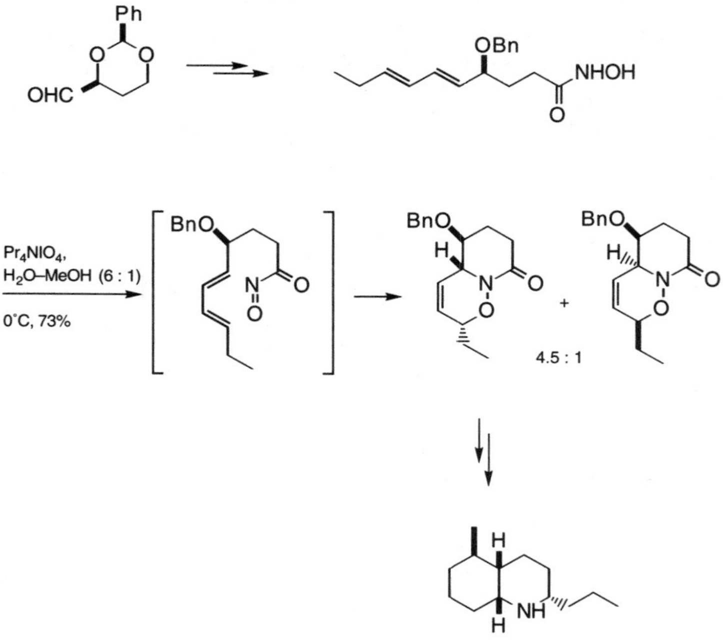

Scheme 2.14

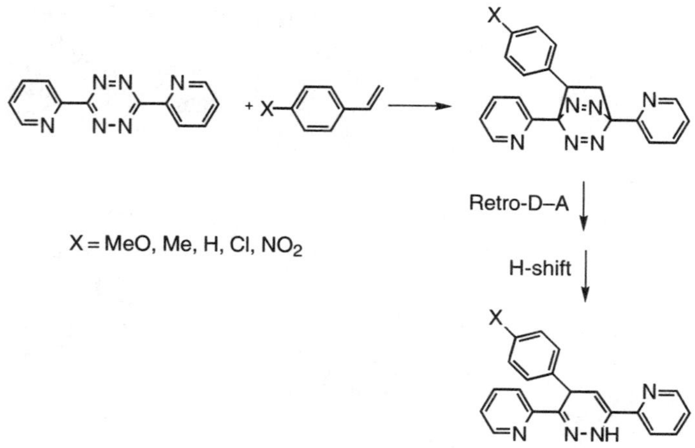

Scheme 2.15

The hetero-Diels–Alder reaction has also utilized dienophiles in which both reactive centers are heteroatoms. Kibayashi reported that the intramolecular hetero-Diels–Alder cycloaddition of chiral acylnitroso compounds showed a pronounced enhancement of the *trans* selectivity in aqueous medium compared with the selectivity in nonaqueous conditions (Scheme 2.13) (68). The reaction was readily applied in the total synthesis of (−)-pumiliotoxin C (Scheme 2.14) (69).

The hetero-Diels–Alder reaction can also employ dienes containing heteroatoms. Cycloaddition of substituted styrenes with di-(2-pyridyl)-1,2,4,5-tetrazine was investigated by Engberts (Scheme 2.15) (70). Again, the rate of the reaction increased dramatically in water-rich media.

2.2 OTHER CYCLIZATION REACTIONS

2.2.1 Alder–Ene Reactions

An ene–iminium one-pot cyclization (Scheme 2.16) proceeds smoothly in a water–THF mixture. The reaction has been used in the asymmetric synthesis of pipecolic acid derivatives (71).

2.2.2 1,3-Dipolar Cycloaddition Reactions

The 1,3-dipolar cyclization of nitrile oxide with dipolarophiles generates structurally important heterocycles [Eq. (2.15)]. As shown by Lee

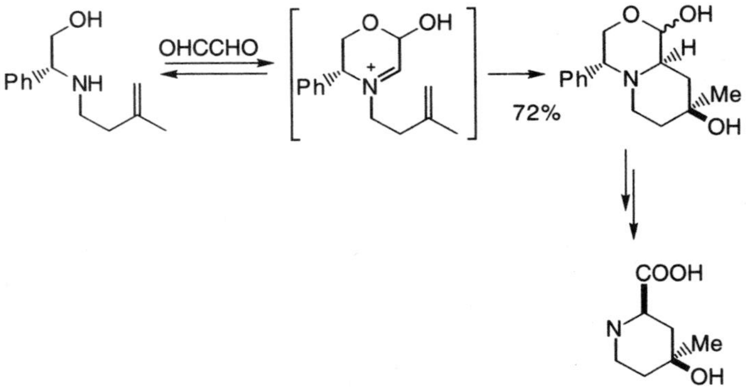

Scheme 2.16

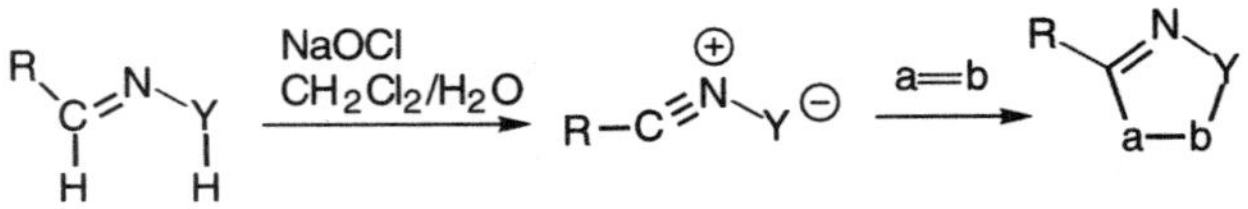

$$\text{(2.15)}$$

(72), the reaction can be carried out in an aqueous-organic biphase system [Eq. (2.16)], in which the nitrile oxide substrates can be generated from oximes or hydrazones *in situ*. The method provides a convenient one-pot procedure for generating a variety of heterocyclic products.

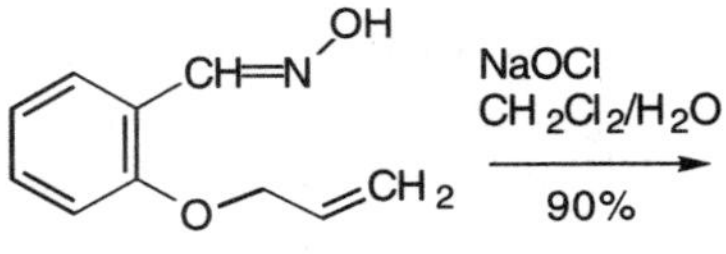

$$\text{(2.16)}$$

Reaction of preformed aromatic nitrile *N*-oxides with alkyl disubstituted benzoquinones gives the 1,3-dipolar cyclization product in aqueous ethanol [Eq. (2.17)] (73). The effect of the polarity of solvents on

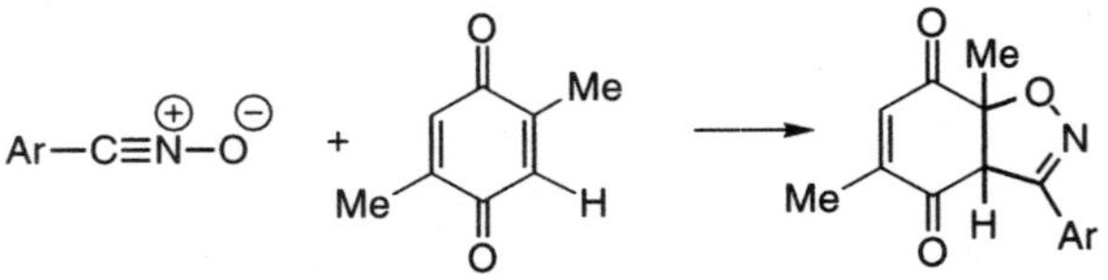

$$\text{(2.17)}$$

the rate of the reaction was investigated. While the reaction is usually slower in more polar solvents than in less polar ones, the use of water as the solvent increases the reactivity. The nitrile oxide reaction has also been catalyzed by baker's yeast (74) and β-cyclodextrin (75).

Reaction of an azomethine ylide, generated *in situ* from methyl *N*-methylglycinate and formaldehyde, with activated olefins generated the corresponding dipolar products (76):

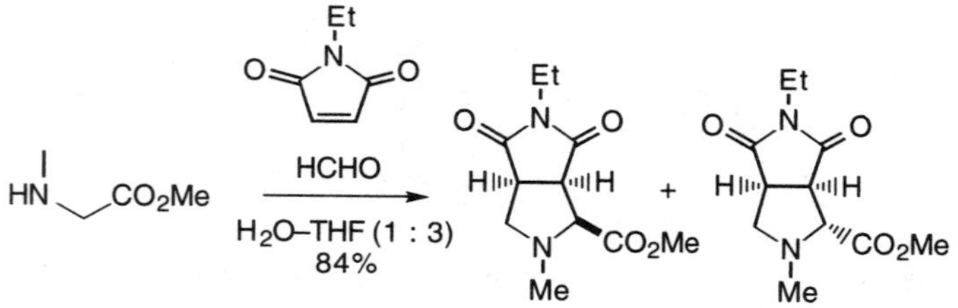

An elegant application of 1,3-dipolar cyclization of an azide derivative in water was reported by De Clercq in the synthesis of (+)-biotin (Scheme 2.17) (77). On thermolysis of the azide compound, a mixture of (+)-biotin and its benzylated derivative was formed directly.

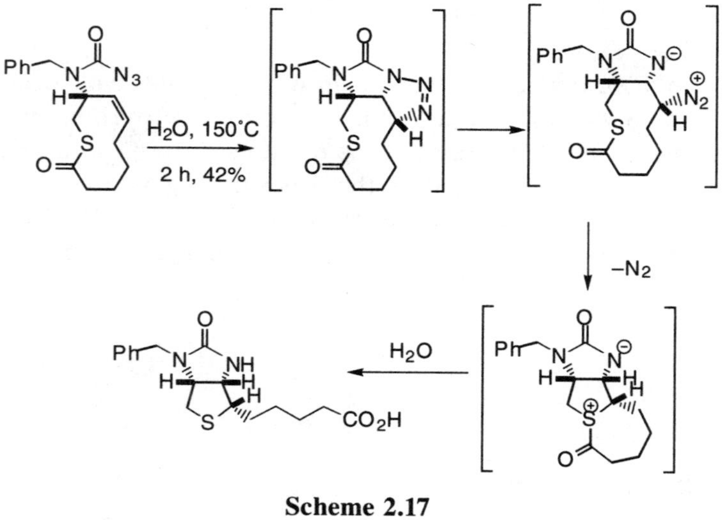

Scheme 2.17

2.2.3 Ozonolysis

The cleavage of alkenes by ozone (ozonolysis) to give carbonyl compounds is a synthetically useful reaction (78). Although there is still disagreement on the exact mechanism, it is generally accepted that two heterocyclic intermediates, 1,2,3-trioxolane and 1,2,4-trioxolane, are involved. The formation of the former could be regarded as a 1,3-dipolar cyclization between ozone and the alkene and the latter due to a retro-cleavage followed by a recombination of the fragments [Eq. (2.19)]. Depending on the synthetic purpose, the 1,2,4-trioxolane can be transformed to alcohols, aldehydes (and ketones), or carboxylic acids.

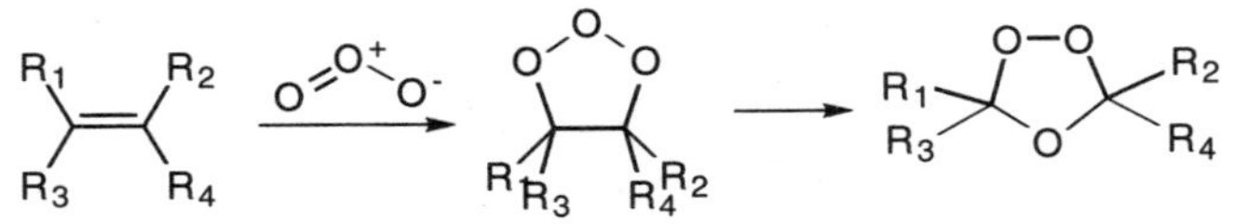

$$(2.19)$$

The reaction is usually performed at low temperatures, and occasionally water has been used as solvent. For example, cyclooctene is ozonized, in the presence of an emulsifier (polyoxyethylated lauryl alcohol), with aqueous alkaline hydrogen peroxide to give α,ω-alkanedicarboxylic acid in one pot (79):

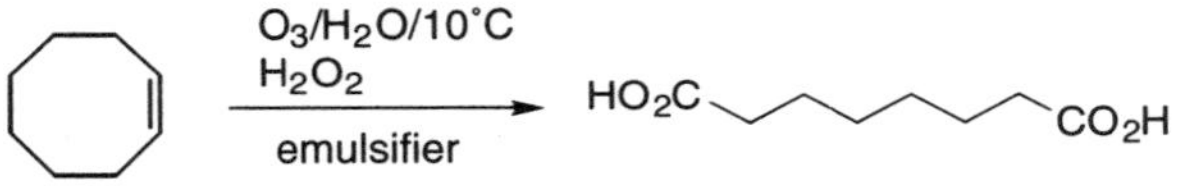

$$(2.20)$$

Ozonolysis of organic compounds in water also has biological interest. Ozone preferentially attacks the base moiety of pyrimidine nucleotides in water (80). For example, the reaction of ozone with uracile in water, having no substitutent at the 1-position, gave the ozonolysis products in Scheme 2.18 (81). The reaction of DNA and RNA with O_3 in an aqueous environment is linked to the damage of biological systems by ozone (82).

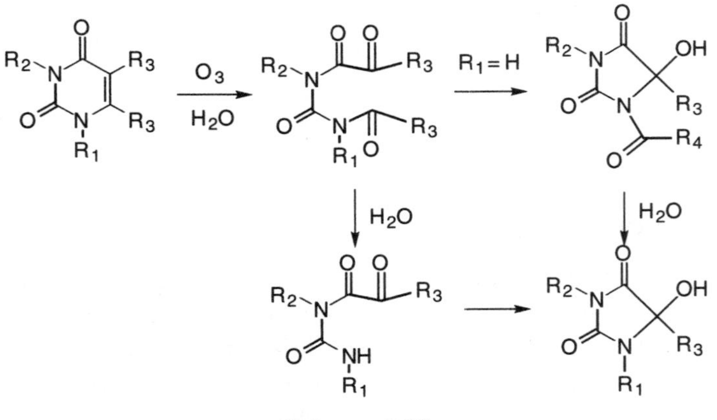

Scheme 2.18

2.3 SIGMATROPIC REARRANGEMENTS

Polar solvents have been known to increase the rate of the Claisen rearrangement reactions (83). Recently, it was observed that Claisen rearrangement reactions were accelerated on going from nonpolar to aqueous solvents (84). For instance, the rearrangements of chorismic acid and related compounds in water were 100 times faster than in methanol [Eq. (2.21)] (85). Because the $\Delta V^{\neq}$ (volume change of activation) of Claisen rearrangements also has a negative value, as in the Diels–Alder reactions, the Claisen rearrangement reaction is expected to be accelerated by water according to the same effect (86, 87).

$$(2.21)$$

Grieco observed a facile [3,3]-sigmatropic rearrangement of an allyl vinyl ether in water, giving rise to an aldehyde [Eq. (2.22)]. The corresponding methyl ester underwent the facile rearrangement similarly. A solvent polarity study on the rearrangement rate of the allyl vinyl ether was conducted in solvent systems ranging from pure methanol to water at 60°C (88). The first-order rate constant for the rearrangement of the allyl vinyl ether in water was 18×10^{-5} s^{-1}, compared to 0.79×10^{-5} s^{-1} in pure methanol.

$$(2.22)$$

The accelerating influence of water as a solvent on the rate of the Claisen rearrangement has also been demonstrated on a number of other substrates. These studies showed that this methodology has potential applications in organic synthesis. For instance, unprotected allyl vinyl ether (**12**), as a 0.01 M solution in water-methanol (2.5:1) containing an equivalent of sodium hydroxide, smoothly underwent rearrangement at ~80°C, affording 85% isolated yield of aldehyde (**13**) in 24 h [Eq. (2.23)] (89). The same rearrangement for the protected analog under organic Claisen reaction conditions had considerable difficulties and often resulted in the elimination of acetaldehyde (90).

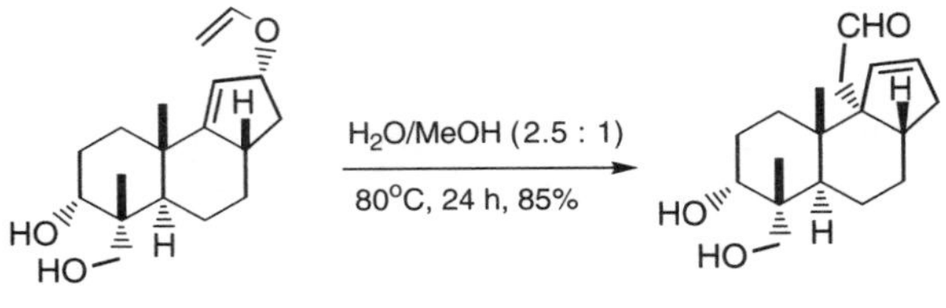

(2.23)

The rearrangement of allyl vinyl ether (**14**) occurred smoothly in water [Eq. (2.24)], while the corresponding methyl ester led to recovered starting material only on prolonged heating in decalin.

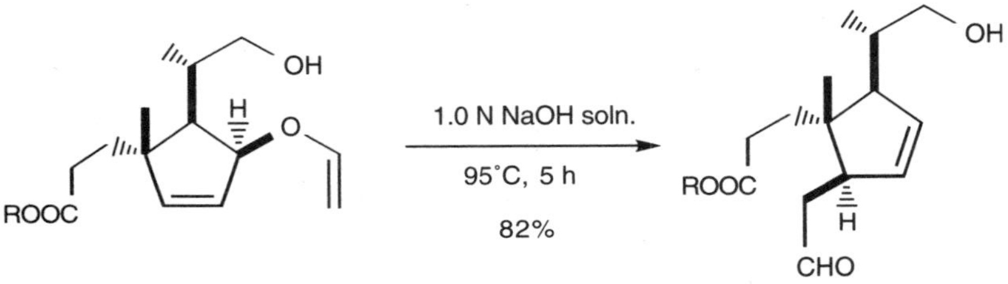

(2.24)

A surprisingly facile rearrangement of fenestrene (**15**) took place in aqueous methanol (3:1) to give fenestrene aldehyde (**16**), which possessed a *trans* ring fusion between the two 5-membered rings [Eq. (2.25)]. Previous attempts to synthesize such a ring fusion and to employ Claisen rearrangements within the carbon framework of a fenestrene system had been unsuccessful.

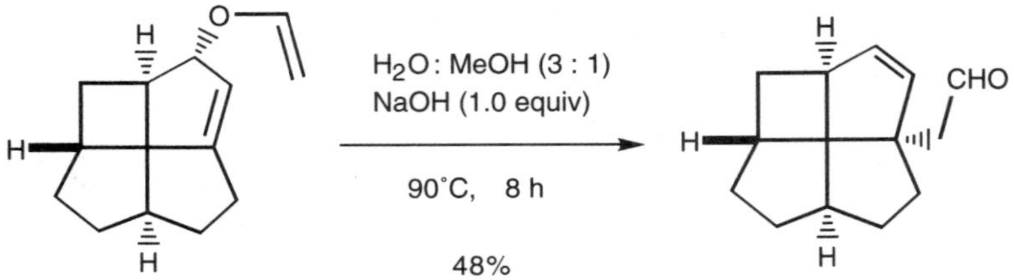

(2.25)

The use of the glucose chiral auxiliary by Lubineau et al. led to moderate asymmetric induction in the Claisen rearrangement of compound **17** (20% d.e.). Compounds **17** and **18** were both water-soluble because of the high polarity of the glucose moiety [Eq. (2.26)] (91). Since it could be removed easily, glucose functioned here as a chiral auxiliary. After separation of the diastereomers, enantiomerically pure substances could be obtained.

17   **18**

(2.26)

The origin of the rate acceleration in Claisen rearrangement is still controversial. A self-consistent-field solvation model was applied to the aqueous medium Claisen rearrangement (92). The aqueous acceleration of the Claisen rearrangement was suggested to be due to solvent-induced polarization and first-hydration-shell hydrophilic effects (93). Theoretical studies by Jorgensen (94) and Gajewski (95) suggested that increased hydrogen bonding in the transition state is responsible for the observed acceleration. On the other hand, studies by Gajewski on the secondary deuterium kinetic isotope effects argue against the involvement of an ionic transition state (96). A combined quantum-mechanical/statistical mechanical approach used by Gao indicated that different substrates have different degree of acceleration (97). Other models have also been used for theoretical study of the rate acceleration (98).

2.4 PHOTOCHEMICAL CYCLOADDITION REACTIONS

An excellent review of organic photochemistry in organized media, including aqueous solvent, has been reported by Ramamurthy (99). The quantum efficiency for photodimerization of thymine, uracil, and their derivatives increased considerably in water in comparison to other organic solvents (Scheme 2.19). The increased quantum efficiency is attributed to the preassociation of reactants at the ground state.

Ramamurthy has shown that organic substrates (having poor solubilities in water), such as stilbenes and alkyl cinnamates, photodimerize

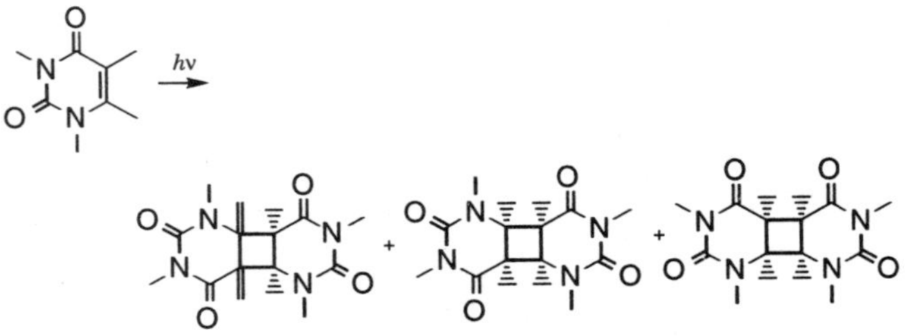

Water	27.8%	63.1%	9.1% $\varnothing = 0.015$
Acetonitrile	24.9%	68.2%	6.7% $\varnothing = 0.0047$
Methanol	31.4%	68.6%	— $\varnothing = 0.004$

Scheme 2.19

efficiently in water [Eq. (2.27)]. The same reaction in organic solvents, such as benzene, leads mainly to *cis–trans* isomerizations (100). As in the case of Diels–Alder reaction, the addition of LiCl (increasing hydrophobic effect) increases the yield of dimerizations. On the other hand, the addition of quanidinium chloride (decreasing the hydrophobic effect) lowers the yield of the product. Similar results were obtained with alkyl cinnamates (101).

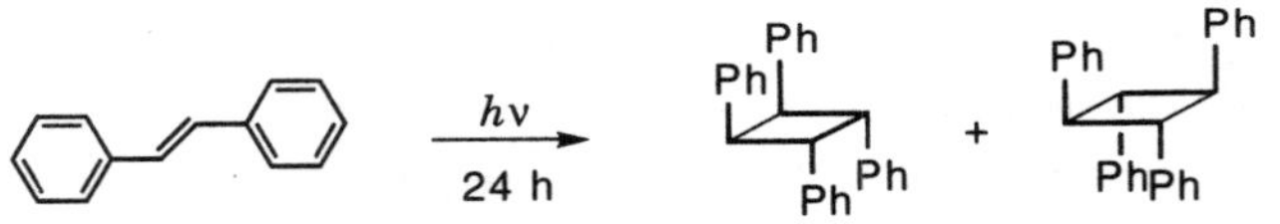

(2.27)

Benzene	0%	0%
Water	12%	10%
Water + LiCl	25%	17%
Water + guanidinium chloride	8%	6%

Similarly, coumarin dimerized more efficiently in water than in organic solvents [Eq. (2.28)] (102). The quantum yield of the dimerization in water is more than 100 times higher than that in benzene and methanol.

When a surfactant is added in water, it will aggregate in forming micelles. The formation of such micelles has also been found to have a significant effect on the regio- and stereoselectivity of photochemical

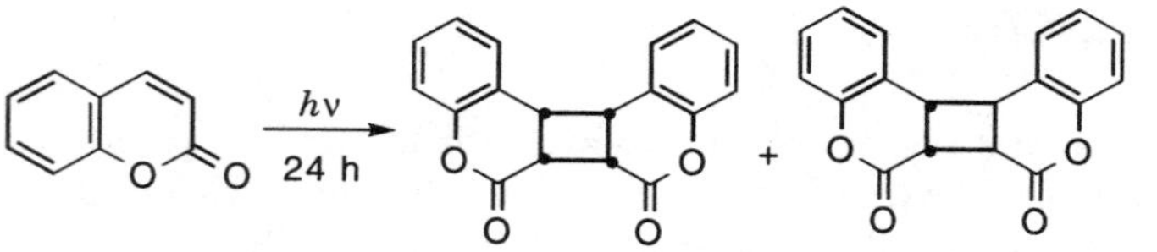

(2.28)

Solvent	Quantum Yield
Benzene	$< 10^{-5}$
Methanol	$< 10^{-5}$
Water	2×10^{-3}

reactions. In the micellar case, the hydrophobic interior of micelles provides a hydrophobic pocket within the bulk water solvent. An analogous situation of hydrophobic cage effect is the use of cyclodextrin. Thus a selectivity in product formation could be expected also in this case. Indeed, while four isomers are generated for the photodimerization of anthracene-2-sulfonate, the same reaction gives only one isomer when β-cyclodextrin is present (Scheme 2.20) (103).

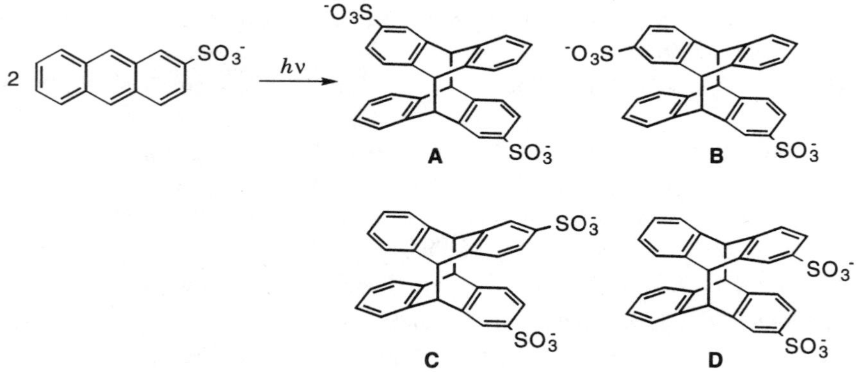

Medium	Product (**A : B : C : D**)
Water	$1 : 0.8 : 0.4 : 0.05$
Water + β-CD	**A** only

Scheme 2.20

In principle, other photochemical electrocyclic reactions should also be susceptible to the influence of water in the same way.

REFERENCES

1. For general reviews of pericyclic reactions, see Woodward, R. B.; Hoffmann, R. *The Conservation of Orbital Symmetry*, Cambridge University Press, London, 1972; Marchand, A. P.; Lehr, R. E. *Pericyclic Reactions*, Academic Press, New York, 1977.

2. van Eldik, R.; Asano, T.; Le Nobel, W. J. *Chem. Rev.* **89**, 549 (1989).

3. Dauben, W. G.; Krabenhaft, H. O. *J. Am. Chem. Soc.* **98**, 1992 (1976).

4. Kumar, A. *J. Org. Chem.* **59**, 230 (1994).

5. Carruthers, W. *Cycloaddition Reactions in Organic Synthesis, Tetrahedron Organic Chemistry Series*, Vol. 8, Baldwin, J. E.; Magnus, P. D., eds., Pergamon Press, 1990.

6. Blake, J. F.; Jorgensen, W. L. *J. Am. Chem. Soc.* **113**, 7430 (1991); Blake, J. F.; Lim, D.; Jorgensen, W. L. *J. Org. Chem.* **59**, 803 (1994).

7. Breslow, R.; Guo, T. *J. Am. Chem. Soc.* **110**, 5613 (1988).

8. Breslow, R.; Rizzo, C. J. *J. Am. Chem. Soc.* **113**, 4340 (1991); for recent theoretical studies on hydrophobic effect, see Muller, N. *Acc. Chem. Res.* **23**, 23 (1990).

9. (a) Schneider, H. J.; Sangwan, N. K. *J. Chem. Soc., Chem. Commun.* 1787 (1986); (b) Schneider, H. J.; Sangwan, N. K. *Angew. Chem., Int. Ed. Engl.* **26**, 896 (1987); Hunt, I.; Johnson, C. D. *J. Chem. Soc. Perkin Trans* 2, 1051 (1991).

10. Sangwan, N. K.; Schneider, H. J. *J. Chem. Soc., Perkin Trans.* 2, 1223 (1989).

11. Blokzijl, W.; Blandamer, M. J.; Engberts, J. B. F. N. *J. Am. Chem. Soc.* **113**, 4241 (1991); Blokzijl, W.; Blandamer, M. J.; Engberts, J. B. F. N. *J. Am. Chem. Soc.* **114**, 5440 (1992); Otto, S.; Blokzijl, W.; Engberts, J. B. F. N. *J. Org. Chem.* **59**, 5372 (1994).

12. Engberts, J. B. F. N. *Pure & Appl. Chem.* **67**, 823 (1995).

13. Jenner, G. *Tetrahedron Lett.* **35**, 1189 (1994).

14. Pai, C. K. Smith, M. B. *J. Org. Chem.* **60**, 3731 (1995); Smith, M. B.; Fay, J. N.; Son, Y. C. *Chem. Lett.* 2451 (1992).

15. Griesbeck, A. G. *Tetrahedron Lett.* **29**, 3477 (1988).

16. Braisted, A. C.; Schultz, P. G. *J. Am. Chem. Soc.* **112**, 7430 (1990); Hilvert, D.; Hill, K. W.; Nared, K.; Auditor, M. T. M. *J. Am. Chem. Soc.* **111**, 9261 (1989); Governeur, V. E.; Houk, K. N.; Pascual-Theresa, B.; Beno, B.; Janda, K. D.; Lerner, R. A. *Science* **262**, 204 (1993).

17. Meekel, A. A. P.; Resmini, M.; Pandit, U. K. *J. Chem. Soc. Chem. Commun.* 571 (1995).

18. Jackson, D. Y.; Jacobs, J. W.; Sugasawara, R.; Reich, S. H.; Bartlett, P. A.; Schulz, P. G. *J. Am. Chem. Soc.* **110**, 4841 (1988); Hilvert, D.; Car-

penter, S. H.; Nared, K. D.; Auditor, M. T. M. *Proc. Natl. Acad. Sci.* (USA) **85**, 4953 (1988).

19. Braisted, A. C.; Schultz, P. G. *J. Am. Chem. Soc.* **116**, 2211 (1994).
20. For reviews, see Breslow, R. *Acc. Chem. Res.* **24**, 159 (1991); Grieco, P. A. *Aldrichim. Acta* **24**, 59 (1991).
21. (a) Diels, O.; Alder, K. *Liebigs Ann. Chem.* **490**, 243 (1931); (b) Hopff, H.; Rautenstrauch, C. W. U.S. Patent 2,262,002; *Chem. Abstr.* **36**, 1046 (1942); (c) Woodward, R. B.; Baer, H. *J. Am. Chem. Soc.* **70**, 1161 (1948); (d) Koch, H.; Kotlan, J.; Markert, H.; *Monatsh. Chem.* **96**, 1646 (1965).
22. Rideout, D. C.; Breslow, R. *J. Am. Chem. Soc.* **102**, 7816 (1980).
23. Ben-Naim, A. *Hydrophobic Interactions*, Plenum Press, New York, 1980; Tanford, C. *The Hydrophobic Effect*, 2nd ed., Wiley, New York, 1980.
24. von Hippel, P. H.; Schleich, T. *Acc. Chem. Res.* **2**, 257 (1969).
25. Breslow, R.; Maitra, U.; Rideout, D. *Tetrahedron Lett.* **24**, 1901 (1983); Breslow, R.; Maitra, U. *Tetrahedron Lett.* **25**, 1239 (1984).
26. Berson, J. A.; Hamlet, Z.; Mueller, W. A. *J. Am. Chem. Soc.* **84**, 297 (1962); Samil, A.A.Z.; de Savignac, A.; Rico, I.; Lattes, A. *Tetrahedron* **41**, 3683 (1985).
27. Sternbach, D. D.; Rossana, D. M. *J. Am. Chem. Soc.* **104**, 5853 (1982).
28. Wang, W. B.; Roskamp, E. J. *Tetrahedron Lett.* **33**, 7631 (1992).
29. Keay, B. A. *J. Chem. Soc., Chem. Commun.* 420 (1987).
30. Bourgeois-Guy, A.; Gore, J. *Bull. Soc. Chim. Fr.* **129**, 490 (1992).
31. Gonzalez, A.; Holt, S. L. *J. Org. Chem.* **47**, 3186 (1982).
32. Elguero J.; Goya, P.; Paez, J. A.; Cativiela, C.; Mayoral, J. A. *Synth. Commun.* **19**, 473 (1989).
33. Grieco, P. A.; Garner, P.; He, Z. M. *Tetrahedron Lett.* **24**, 1897 (1983).
34. Grieco, P. A.; Yoshida, K.; Garner, P. *J. Org. Chem.* **48**, 3137 (1983).
35. Grieco, P. A.; Yoshida, K.; He, Z. M. *Tetrahedron Lett.* **25**, 5715 (1984).
36. Grieco, P. A.; Galatsis, P.; Spohn, R. F. *Tetrahedron* **42**, 2847 (1986).
37. Proust, S. M.; Ridley, D. D. *Aust. J. Chem.* **37**, 1677 (1984).
38. Grieco, P. A.; Garner, P.; Yoshida, K.; Huffmann, J. C. *Tetrahedron Lett.* **24**, 3807 (1983).
39. Yoshida, K.; Grieco, P. A. *J. Org. Chem.* **49**, 5257 (1984).
40. Drewes, S. E. ; Grieco, P. A.; Huffman, J. C. *J. Org. Chem.* **50**, 1309 (1985).
41. Brandes, E.; Grieco, P. A.; Garner, P. *J. Chem. Soc., Chem. Commun.* 500 (1988).
42. Yoshida, K.; Grieco, P. A. *Chem. Lett.* 155 (1985).
43. Williams, D. R.; Gaston, R. D.; Horton, I. B., III *Tetrahedron Lett.* **26**, 1391 (1985).

44. Van Royen, L. A.; Mijngheer, R.; Declercq, P. J. *Tetrahedron* **41**, 4667 (1985).

45. (a) Grootaert, W. M.; Declercq, P. J. *Tetrahedron Lett.* **27**, 1731 (1986); (b) Nuyttens, F.; Appendino, G.; De Clercq, P. J. *Synlett.* 526 (1991).

46. (a) Lubineau, A.; Queneau, Y. *Tetrahedron Lett.* **26**, 2653 (1985); (b) Lubineau, A.; Queneau, Y. *J. Org. Chem.* **52**, 1001 (1987); (c) Lubineau, A.; Queneau, Y. *Tetrahedron* **45**, 6697 (1989); (d) Lubineau, A.; Bienayme, H.; Queneau, Y.; Scherrmann, M. C. *New J. Chem.* **18**, 279 (1994).

47. Zezza, C. A.; Smith, M. B. *J. Org. Chem.* **53**, 1161 (1988).

48. Saksena, A. K.; Girijavallabhan, V. M.; Chen, Y. T.; Jao, E.; Pike, R. E.; Desai, J. A.; Rane, D.; Ganguly, A. K. *Heterocycles* **35**, 129 (1993).

49. Hollis, T. K.; Robinson, N. P.; Bosnich, B. *J. Am. Chem. Soc.* **114**, 5464 (1992).

50. Kobayashi, S.; Hachiya, I.; Araki, M.; Ishitani, H. *Tetrahedron Lett.* **34**, 3755 (1993).

51. Otto, S.; Engberts, J. B. F. N. *Tetrahedron Lett.* **36**, 2645 (1995).

52. Colonna, S.; Manfredi, A.; Annunziata, R. *Tetrahedron Lett.* **29**, 3347 (1988).

53. Cativiela, C.; Diaz de Villegas, M. D.; Mayoral, J. A.; Avenoza, A.; Peregrina, J. M. *Tetrahedron* **49**, 677 (1993).

54. Boger, D.; Weireb, S. *Hetero-Diels–Alder Methodology in Organic Synthesis*, Academic Press, Orlando, 1987.

55. Grieco, P. A.; Larsen, S. D. *J. Am. Chem. Soc.* **107**, 1768 (1985).

56. Oppolzer, W. *Angew. Chem., Int. Ed. Engl.* **11**, 1031 (1972).

57. Grieco, P. A.; Larsen, S. D.; Fobare, W. F. *Tetrahedron Lett.* **27**, 1975 (1986).

58. Grieco, P. A.; Parker, D. T.; Fobare, W. F.; Ruckle, R. *J. Am. Chem. Soc.* **109**, 5859 (1987).

59. Grieco, P. A.; Bahsas, A. *J. Org. Chem.* **52**, 5746 (1987).

60. (a) Waldmann, H. *Angew. Chem.* **100**, 307 (1988); *Angew. Chem. Int. Ed. Engl.* **27**, 274 (1988); (b) Waldmann, H. *Liebigs Ann. Chem.* 231 (1989); (c) Waldmann, H; Braun, M. *Liebigs Ann. Chem.* 1045 (1991).

61. Lock, R.; Waldmann, H. *Tetrahedron Lett.* **37**, 2753 (1996).

62. Yu, L.; Chen, D.; Wang, P. G *Tetrahedon Lett.* **37**, 2169 (1996).

63. Lubineau, A.; Auge, J.; Lubin, N. *Tetrahedron Lett.* **32**, 7529 (1991).

64. Grieco, P. A.; Henry, K. J.; Nunes, J. J.; Matt, J. E., Jr. *J. Chem. Soc., Chem. Commun.* 368 (1992).

65. MacKeith, R. A.; McCague, R.; Olivo, H. F.; Palmer, C. F.; Roberts, S. M. *J. Chem. Soc., Perkin Trans. 1* 313 (1993).

66. McCague, R.; Olivo, H. F.; Roberts, S. M. *Tetrahedron Lett.* **34**, 313 (1993).

67. Lubineau, A.; Auge, J.; Lubin, N. *Tetrahedron* **49**, 4639 (1993); Lubineau, A.; Queneau, Y. *J. Carbohydrate Chem.* **14**, 1295 (1995).

68. Naruse, M.; Aoyagi, S.; Kibayashi, C. *Tetrahedron Lett.* **35**, 595 (1994).

69. Naruse, M.; Aoyagi, S.; Kibayashi, C. *Tetrahedron Lett.* **35**, 9213 (1994).

70. Wijnen, J. W.; Zavarise, S.; Engberts, J. B. F. N. *J. Org. Chem.* **61**, 2001 (1996).

71. Agami, C.; Couty, F.; Poursoulis, M.; Vaissermann, J. *Tetrahedron* **48**, 431 (1992).

72. Lee, G. A. *Synthesis* 508 (1982).

73. Inoue, Y.; Araki, K.; Shiraishi, S. *Bull. Chem. Soc. Jpn.* **64**, 3079 (1991).

74. Rao, K. R.; Bhanumathi, N.; Srinivasan, T. N.; Sattur, P. B. *Tetrahedron Lett.* **31**, 899 (1990).

75. Rao, K. R.; Bhanumathi, N.; Sattur, P. B. *Tetrahedron Lett.* **31**, 3201 (1990).

76. Lubineau, A.; Bouchain, G.; Queneau, Y. *J. Chem. Soc., Perkin Trans. 1* 2433 (1995).

77. Deroose, F. D.; De Clercq, P. J. *Tetrahedron Lett.* **35**, 2615 (1994).

78. For some general reviews, see Bailey, P. S. *Chem. Rev.* **58**, 925 (1958); Criegee, R. *Rec. Chem. Progr.* **18**, 111 (1957).

79. Fremery, M. I.; Fields, E. K. *J. Org. Chem.* **28**, 2537 (1963).

80. Ishizaki, K.; Shinriki, N.; Ikehata, A.; Ueda, T. *Chem. Pharm. Bull.* **29**, 868 (1981).

81. Matsui, M.; Nakazumi, H.; Kamiya, K.; Yatome, C.; Shibata, K.; Muramatsu, H. *Chem. Lett.* 723 (1989).

82. John, T. B.; Flint, R. B. *J. Am. Chem. Soc.* **53**, 1082 (1931); Kolonko, K. J.; Shapiro, R. H.; Barkley, R. M.; Sievers, R. E. *J. Org. Chem.* **44**, 3769 (1979).

83. White, W. N.; Wolfarth, E. F. *J. Org. Chem.* **35**, 2196, 3585 (1970).

84. (a) Ponaras, A. A. *J. Org. Chem.* **48**, 3866 (1983); (b) Coates, R. M.; Rogers, B. D.; Hobbs, S. J.; Peck, D. R.; Curran, D. P. *J. Am. Chem. Soc.* **109**, 1160 (1987); (c) Gajewski, J. J.; Jurayj, J.;.Kimbrough, D. R.; Gande, M. E.; Ganem, B. Carpenter, B. K. *J. Am. Chem. Soc.* **109**, 1170 (1987).

85. Copley, S. D.; Knowles, J. *J. Am. Chem. Soc.* **109**, 5008 (1987).

86. Brower, K. R. *J. Am. Chem. Soc.* **83**, 4370 (1961).

87. Walling, C.; Naiman, M. *J. Am. Chem. Soc.* **84**, 2628 (1962).

88. Brandes, E. ; Grieco, P. A.; Gajewski, J. J. *J. Org. Chem.* **54**, 515 (1989).

89. Grieco, P. A.; Brandes, E. B.; McCann, S.; Clark, J. D. *J. Org. Chem.* **54**, 5849 (1989).

90. McMurry, J. E.; Andrus, A.; Ksander, G. M. ; Musser, J. H.; Johnson, M. A. *Tetrahedron* **37**, 319 (Suppl. 1) (1981).

91. Lubineau, A.; Auge, J.; Bellanger, N.; Caillebourdin, S. *Tetrahedron Lett.* **31**, 4147 (1990); *J. Chem. Soc. Perkin Trans.* **1**, 1631 (1992).

92. Cramer, C. J.; Truhlar, D. G., *J. Am. Chem. Soc.* **114**, 8794 (1992).

93. Gao, J. *J. Am. Chem. Soc.* **116**, 1563 (1994); Davidson, M. M.; Hillier, I. H.; Hall, R. J.; Burton, N. A. *J. Am. Chem. Soc.* **116**, 9294 (1994).

94. Severance, D. L.; Jorgensen, W. L. *J. Am. Chem. Soc.* **114**, 10966 (1992).

95. Gajewski, J. J. *J. Org. Chem.* **57**, 5500 (1992).

96. Gajewski, J. J.; Brichford, N. L. *J. Am. Chem. Soc.* **116**, 3165 (1994).

97. Sehgal, A.; Shao, L.; Gao, J. *J. Am. Chem. Soc.* **117**, 11337 (1995).

98. Davidson, M. M.; Hillier, I. H. *J. Phys. Chem.* **99**, 6748 (1995); Davidson, M. M.; Hillier, I. H.; Vincent, M. A. *Chem. Physc. Lett.* **246**, 536 (1995).

99. Ramamurthy, R. *Tetrahedron* **42**, 5753 (1986).

100. Sayamala, M. S.; Ramamurthy, V. *J. Org. Chem.* **51**, 3712 (1986).

101. Ito, Y.; Kajita, T.; Kunimoto, K.; Matsuura, T. *J. Org. Chem.* **54**, 587 (1989).

102. Muthuramu, K.; Ramamurthy, V. *J. Org. Chem.* **47**, 3976 (1982).

103. Tamaki, T. *Chem. Lett.* 1984, 53; Tamaki, T.; Kokubu, T. *J. Inclusion Phenom.* **2**, 815 (1984).

CHAPTER 3

NUCLEOPHILIC ADDITIONS
AND SUBSTITUTIONS

It is possible not to perceive what one knows and it is possible to perceive.

—Socrates (470–399 B.C.)*

Nucleophiles are molecules or ions with nonbonded pairs of electrons, with which they can form one or more new bonds. These species are electron-rich and can react with electrophiles (electron-poor ions or molecules). The electrophiles can have either positive charge or center of low electron density. When the electrophile is a polarized unsaturated bond, the reaction leads to a nucleophilic addition product. In the case of nucleophilic addition involving anionic nucleophiles, the reaction often can be catalyzed by either bases or acids. The activation of acids and bases on these reactions are due to different effects. While the use of a base generates or increases the effective concentration of the reactive nucleophile from the corresponding pronucleophile, an acid catalyzes the reaction through complexation with the polar group of the reactant, further polarizing the electrophile and lowering the activation energy of the reaction. Thus, the use of neutral nucleophiles for additions will be catalyzed only by acids. Synthetically, the nucleophilic

*N. R. Gulley, *The Philosophy of Socrates*, Macmillan, London; St. Martin's Press, New York, 1968.

addition of unsaturated bonds with carbon-based nucleophiles is the most useful. This chapter will focus on these reactions.

3.1 NUCLEOPHILIC ADDITION INVOLVING NUCLEOPHILES WITH ACIDIC HYDROGENS

The reaction of activated methyl and methylene compounds with a variety of electrophiles represents a large group of important classic reactions for carbon–carbon bond formation. Among these useful reactions are Michael addition, aldol-type reaction, Wittig reaction, Darzen's reaction, and Mannich-type reaction. Very often, these reactions can be carried out in an organic–aqueous two-phase system, with the use of an aqueous basic solution (e.g., aqueous NaOH). Since the 1970s, the use of phase-transfer catalysts greatly improved the scope and efficiency of these reactions (1). Recent studies suggested that by using water alone as solvent, these reactions can proceed under more neutral conditions. Since these reactions generally also have negative values of volume of activation (2), the cause of activation of these reactions by water could be the same as in the case of pericyclic reactions: its high internal pressure through hydrophobic effect.

3.1.1 Michael Addition

In the 1970s, Hajos (3) and Wiechert (4) independently reported that the Michael addition of 2-methylcyclopentane-1,3-dione to vinyl ketone in water gives the corresponding conjugated addition product without the use of a basic (pH > 7) catalyst (Scheme 3.1). The Michael addition product further cyclizes to give a 5–6 fused-ring system. The use of water as solvent is much more superior both in yield and in purity of

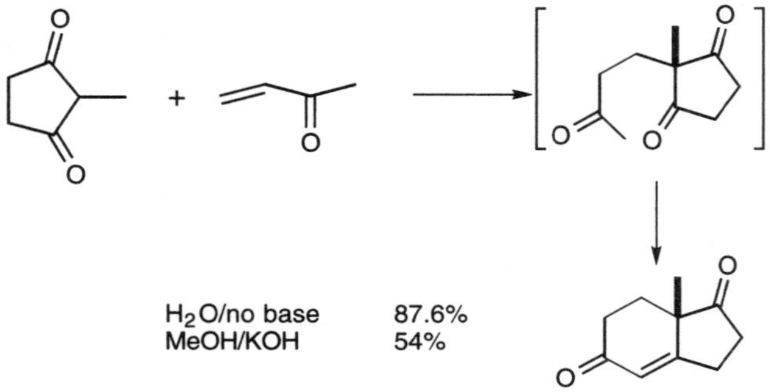

H₂O/no base	87.6%
MeOH/KOH	54%

Scheme 3.1

the product in comparison with the same reaction in methanol with base.

A similar enhancement of reactivity was found in the Michael addition of 2-methyl-cyclohexane-1,3-dione to vinyl ketone, which eventually led to optically pure Wieland–Miescher ketone (Scheme 3.2) (5). The reaction, however, proceeds under more drastic conditions.

Recently, Deslongchamps extended the aqueous Michael addition to acrolein (6). This study has been applied to the total synthesis of 13-α-methyl-14α-hydroxysteroid (Scheme 3.3). The addition of ytterbium

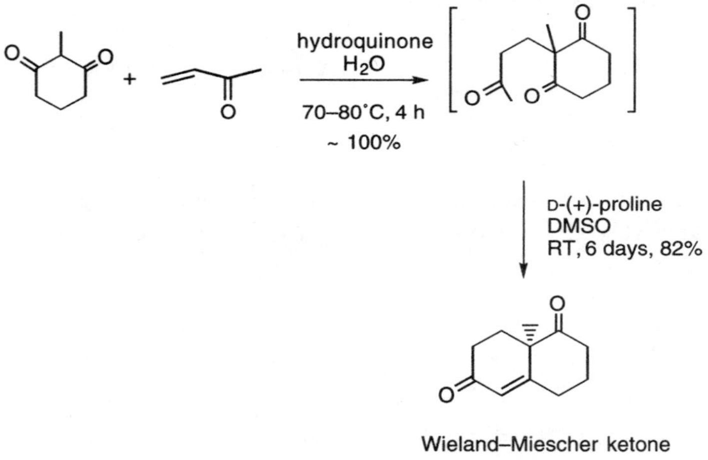

Wieland–Miescher ketone

Scheme 3.2

13-α-methyl-14α-hydroxysteroid

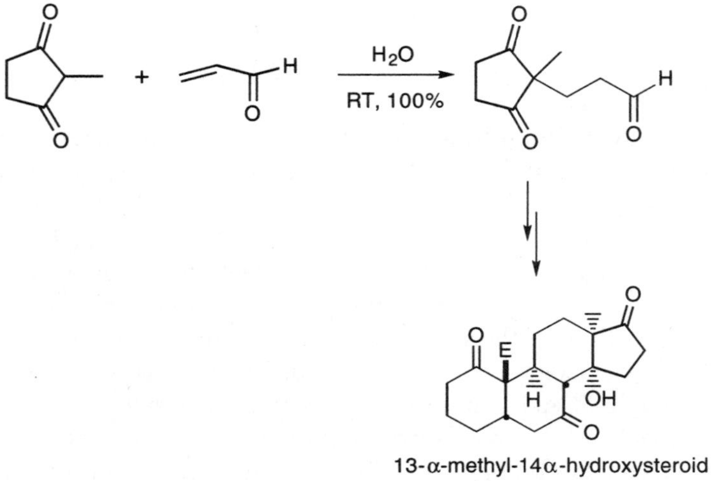

Scheme 3.3

triflate [Yb(OTf)$_3$] further enhances the rate of the Michael addition reactions in water (7).

A significant acceleration of Michael addition was also reported recently by Lubineau in the reaction of nitroalkanes with buten-2-one when going from nonpolar organic solvents to water [Eq. (3.1)] (8). Additives, such as glucose and saccharose, further increase the rate of the reaction.

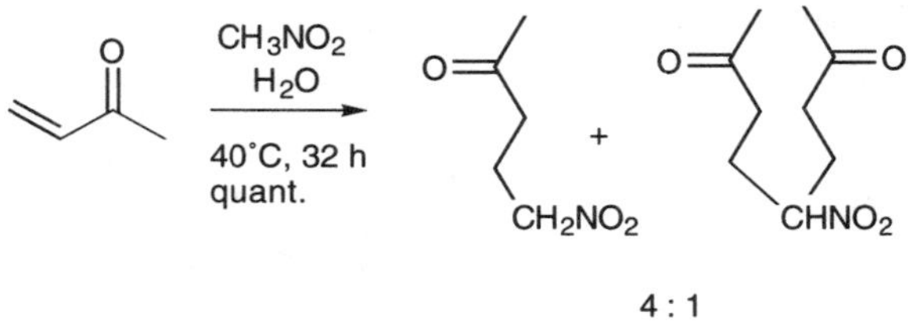

$$4:1 \tag{3.1}$$

Michael addition reactions between ascorbic acid and acrylic enones were effectively carried out in water in the presence of an inorganic acid, rather than a base, as catalyst (9):

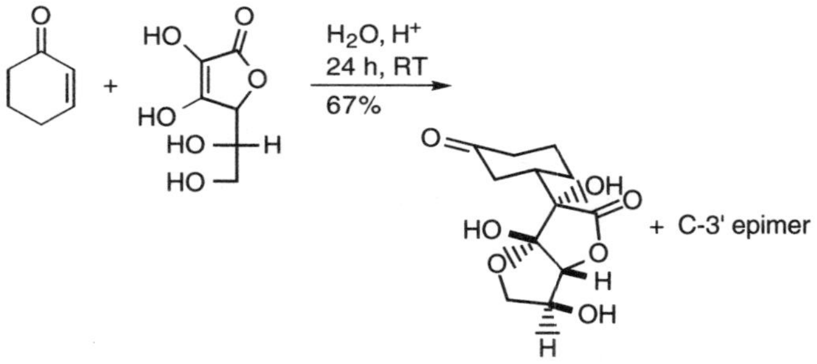

$$+ \text{C-3' epimer} \tag{3.2}$$

The Michael addition in aqueous medium involving heteroatoms as nucleophiles has also been investigated. For example, water-soluble phosphines [e.g., 3,3′,3″-phosphinidynetris(benzenesulfonic acid)] react with α,β-unsaturated acids in water to give the corresponding phosphonium salts quantitatively (10). Similar reactions occur on activated alkynes, leading to vinylphosphonium salts, vinylphosphine oxides, or alkenes depending on the pH of the solution (Scheme 3.4). The phosphines can also isomerize *cis*-electron-deficient olefins to the *trans*-isomer through an addition–elimination process (Scheme 3.5) (11).

For Michael addition with nitrogen nucleophiles, a quantitative study of the Michael addition of activated olefins by using substituted pyridines as nitrogen nucleophiles in water was also reported (12). In this re-

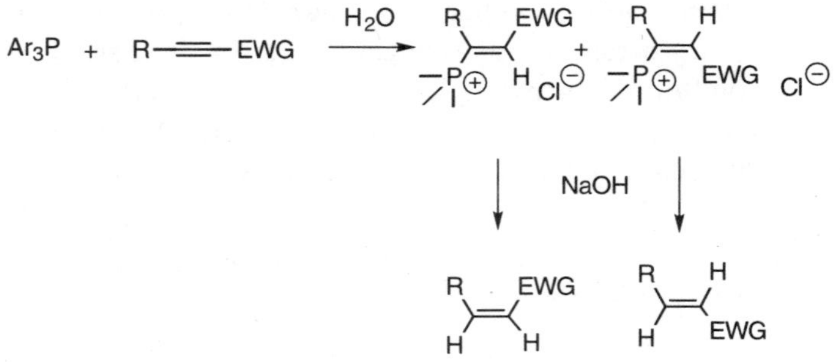

$$Ar_3P \ + \ R-\!\!\!\equiv\!\!\!-EWG \ \xrightarrow{H_2O} \ \ldots$$

NaOH

$PAr_3 = Ph_2P(m\text{-}C_6H_4SO_3Na),\ P(m\text{-}C_6H_4SO_3Na)_3$
$EWG = CHO,\ CO_2R,\ COR$

Scheme 3.4

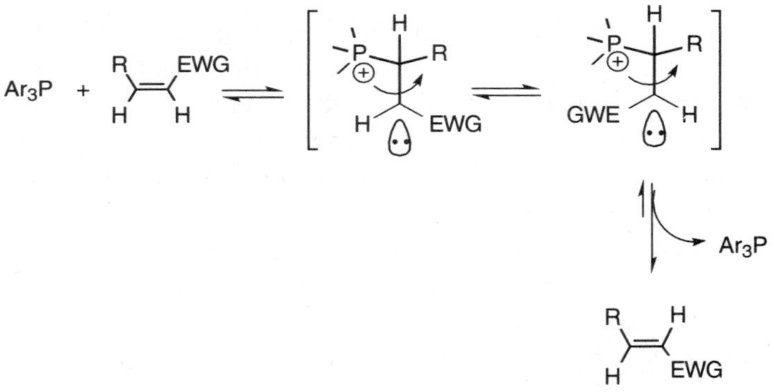

$$Ar_3P \ + \ \ldots \ \rightleftharpoons \ \ldots \ \rightleftharpoons \ \ldots$$

Ar₃P

Scheme 3.5

port, the rate-determining step was investigated. A related reaction between activated olefins with aldehydes in the presence of tertiary amines, the so-called Baylis–Hillman reaction, generates synthetically useful allyl alcohols (13). In some cases, aqueous medium was used for the reaction. The reaction, however, is generally very slow, requiring several days for completion. Recently, Augé et al. studied the reaction in aqueous media in detail (14). A significant increase in reactivity was observed when the reaction is carried out in water [Eq. (3.3)]. The addition of lithium or sodium iodide further increased the reactivity.

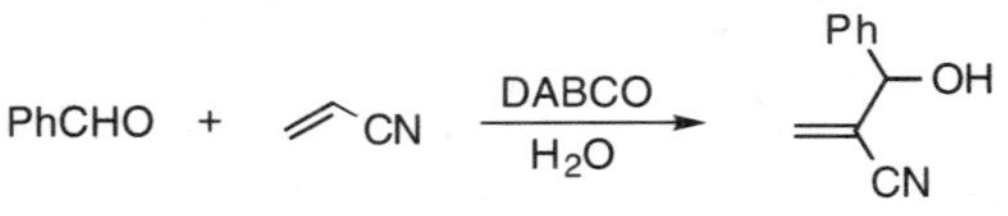

$$PhCHO \ + \ \diagup\!\!\!\diagdown CN \ \xrightarrow[H_2O]{DABCO} \ \ldots \qquad (3.3)$$

The hydrocyanation of conjugated carbonyl compounds is a related reaction (15). Very often such a conjugated addition is carried out under an aqueous condition. For example, in the pioneer work of Lapworth, hydrocyanation of activated olefins was carried out with KCN or (NaCN) in aqueous ethanol in the presence of acetic acid (16):

$$PhCH{=}C(CO_2C_2H_5)_2 \xrightarrow[\substack{4°C}]{\substack{NaCN-AcOH \\ aq\ EtOH}} \underset{\underset{93-96\%}{CN}}{Ph\overset{|}{C}HCH(CO_2C_2H_5)_2} \qquad (3.4)$$

3.1.2 Aldol Reaction

The aldol reaction is important for creating carbon–carbon bonds. The condensation reactions of active methylene compounds such as acetophenone or cyclohexanone with aryl aldehydes gave good yields of aldols along with the dehydration compounds in water (17). The presence of surfactants led mainly to the dehydration reactions. The choice of solvent generally depends on the solubility of reactants and catalyst. The most common solvents for aldol reactions are ethanol, aqueous ethanol, and water (18). The aqueous sodium hydroxide–ether two-phase system has been found to be excellent for the condensation reactions of reactive aliphatic aldehydes (19). By the condensation of an arylaldehyde in alkaline aqueous medium with an arylmethylketone followed by oxidation with hydrogen peroxide, 7- and 3′,4′-substituted flavonols were synthesized under one-pot conditions (20):

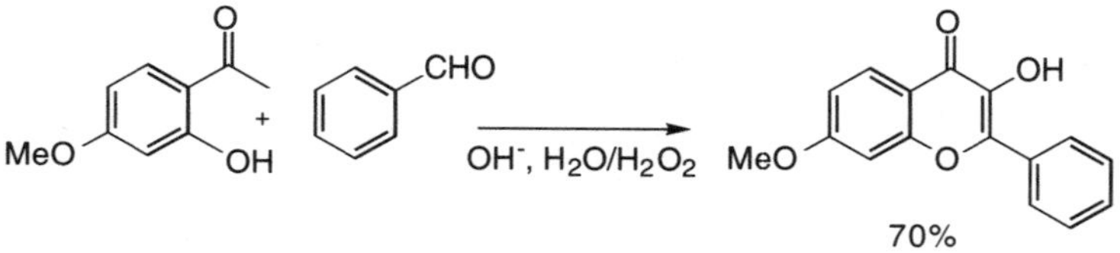

$$\text{(3.5)}$$

Recently, the cross-aldol reaction of silyl enol ethers with carbonyl compounds (the Mukaiyama reaction) was carried out in aqueous solvents without any acid catalyst (21). However, the reaction took several days to complete [Eq. (3.6)]. Presumably, water serves here as a weak

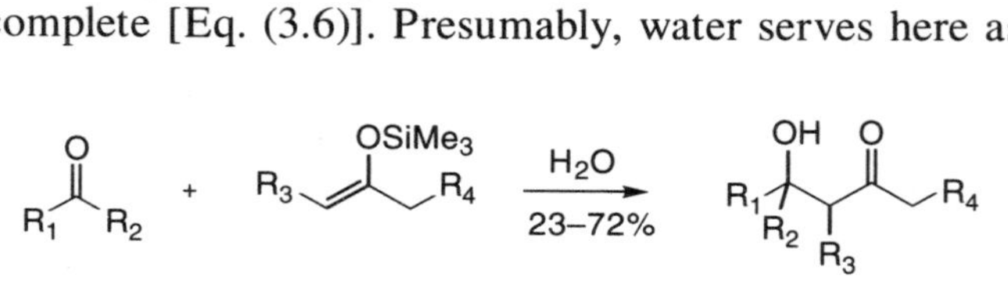

$$\text{(3.6)}$$

Lewis acid. The cross-aldol products showed only a slight *syn* diastereoselectivity, which was the same as when this reaction was carried out in organic solvent under high pressure.

The addition of a catalytic amount of lanthanide triflate (a stronger Lewis acid) greatly improved the rate and the yield of such reactions [Eq. (3.7)] (22). Among the lanthanide triflates, ytterbium triflate [Yb(OTf)$_3$], gadolinium triflate [Gd(OTf)$_3$], and lutetium triflate [Lu(OTf)$_3$] generally gave better yields of the aldol condensation product; the diastereoselectivities of these reactions were moderate. Water-soluble aldehydes were applicable, and the catalyst could be recovered and reused in this procedure.

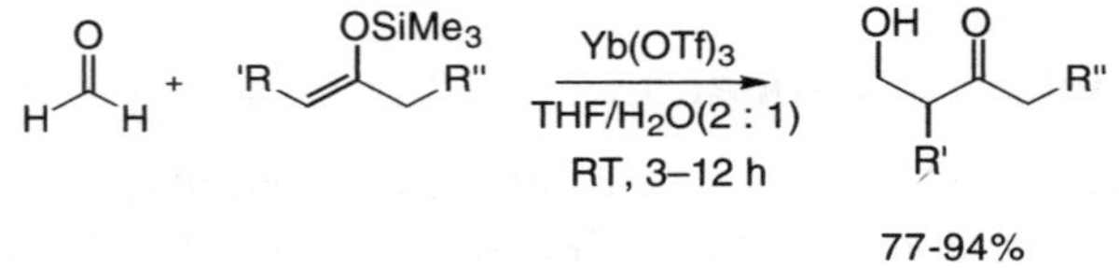

$$(3.7)$$

77-94%

In the presence of complexes of zinc with aminoesters or aminoalcohols, the dehydration can be avoided, and the aldol addition becomes essentially quantitative (23):

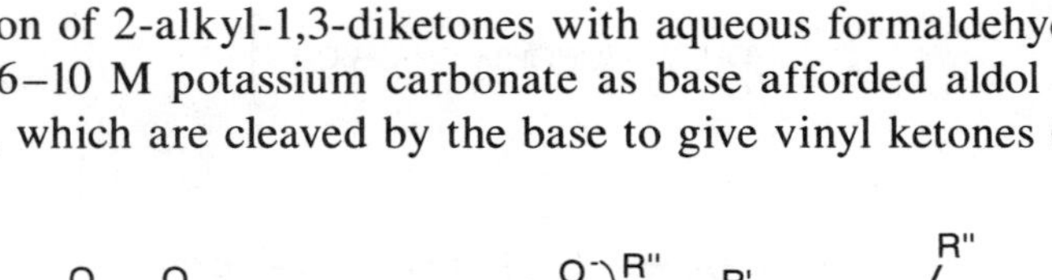

$$(3.8)$$

$Ar = p\text{-}O_2NC_6H_4$

Reaction of 2-alkyl-1,3-diketones with aqueous formaldehyde using aqueous 6–10 M potassium carbonate as base afforded aldol reaction products, which are cleaved by the base to give vinyl ketones (24):

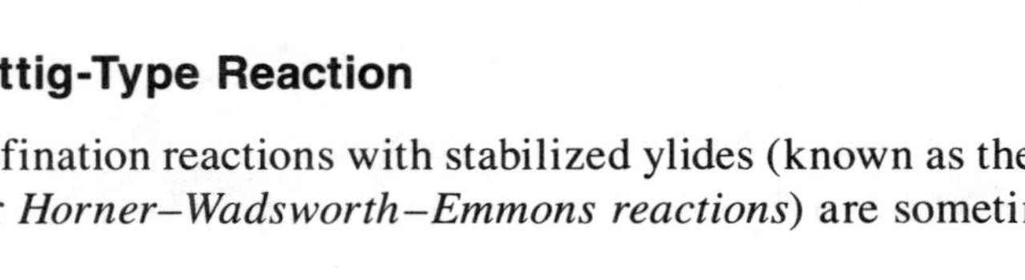

$$(3.9)$$

3.1.3 Wittig-Type Reaction

Wittig olefination reactions with stabilized ylides (known as the *Wittig–Horner* or *Horner–Wadsworth–Emmons reactions*) are sometimes performed in an organic/water biphase system (25). Very often, a phase-transfer catalyst is used. Recently, the use of water alone as solvent has been investigated (26). The reaction proceeds smoothly with a much weaker base, such as K_2CO_3 or $KHCO_3$. No phase-transfer catalyst is required. Compounds with base- and acid-sensitive functional groups can be used directly. For example, under such a condition, β-dimethyl-hydrazoneacetaldehyde can be olefinated efficiently (27):

$$
\underset{\overset{|}{H}}{-N}\overset{N=}{\diagdown}\!\!\!=\!O \quad \xrightarrow[\substack{K_2CO_3/H_2O \\ \text{reflux, 1 h 85\%}}]{(EtO)_2POCH_2CO_2Et} \quad \underset{\overset{|}{H}}{-N}\overset{N=}{\diagdown}\underset{EtO}{\diagdown}\!\!=\!O \qquad (3.10)
$$

3.1.4 Cyanohydrin Formation

Addition reactions of HCN to carbon hetero multiple bonds, such as cyanohydrin formation reactions (28), and $C=N$ or $C\equiv N$ addition reactions (29), have all been performed in aqueous media. In particular, synthesis of higher sugars through cyanohydrin formation, the Kiliani–Fischer synthesis, is a classic reaction and probably is the earliest carbohydrate chemical synthesis without the use of a protecting group (30). The reaction is governed by the acidity of the media. The addition of cyanide to simple aldoses is essentially quantitative at pH 9.1,

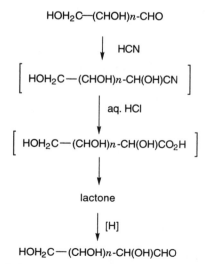

Scheme 3.6   Kiliani–Fischer higher-sugar synthesis.

whereas the reaction is much slower at a lower pH. The cyanohydrins are not isolated but are converted to the corresponding lactones. Reduction of the lactones by sodium amalgam, catalytic hydrogenation, or reduction with sodium borohydride at pH 3–4 in aqueous solution generates the higher aldoses (Scheme 3.6). By controlling the counterion of cyanide, one can change the proportions of diastereomers, originating from the creation of new stereogenic center. The synthesis of, essentially, a single diastereomer is also possible [Eq. (3.11)] (31).

D-Mannose   Single diastereomer

$$(3.11)$$

3.1.5 Benzoin Condensation

The benzoin condensation reaction consists of treating an aromatic aldehyde with potassium cyanide or sodium cyanide, usually in an aqueous ethanolic solution. Recently, Breslow studied the effects of inorganic salts on the rate of the cyanide-catalyzed benzoin condensation in aqueous media [Eq. (3.12)] (32). The reaction is 200 times faster in water than in ethanol. Through the use of a quantitative antihydrophobic effect as probes for transition-state structures, it was postulated that the acceleration of benzoin condensation in water was related to the amount of hydrophobic surfaces that are solvent-accessible in the transition states compared with the initial state (33). Thus the addition of salts, increasing the hydrophobic effect, further increased the rate of the reaction. The addition of γ-cyclodextrin (in which both substrates can fit) also accelerates the reaction, whereas the addition of β-cyclodextrin (with a smaller cavity) inhibits the condensation:

$$(3.12)$$

Salt (5.0 M)	$10^4 \times k_2$ (M^{-1} min^{-1})	Salt (5.0 M)	$10^4 \times k_2$ (M^{-1} min^{-1})
No salt	123	CsCl	153
LiCl	470	CsI	228
LiBr	166	β-CD, 5 mM	45
LiClO$_4$	36	γ-CD, 5 mM	217
LiI	24	KCl	351

The cyanide-catalyzed benzoin condensation can be compared with the benzoin condensation catalyzed by thiamine in the biological systems, elucidated by Breslow (34). While the cyanide-catalyzed benzoin condensation proceeds through a carbanion, stabilized by the cyano group, the thiamine-catalyzed benzoin condensation generated a carbanion stabilized by the thiazolium group (Scheme 3.7). Recently, Breslow has prepared several γ-cyclodextrin thiazolium salts. These cyclodextrin derivatives mimic the action of thiamine and catalyze the benzoin condensation very effectively (35).

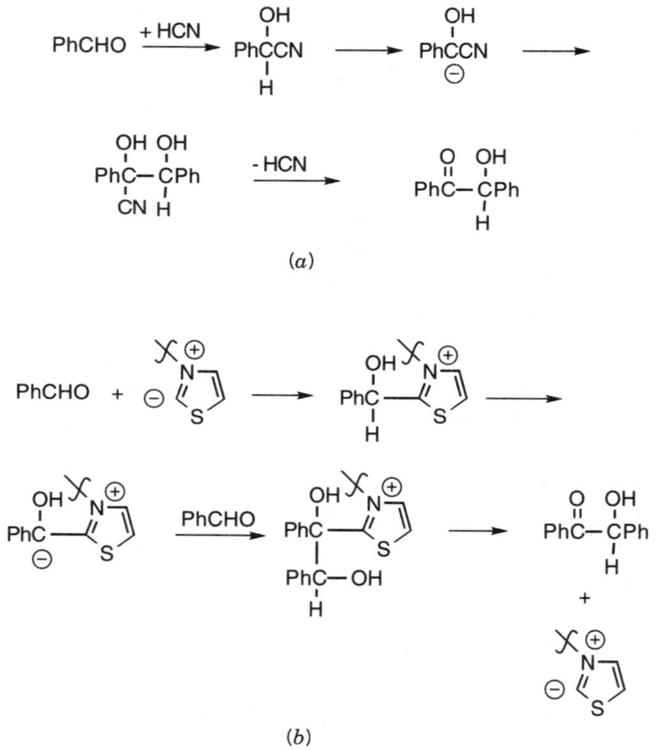

(a)

(b)

Scheme 3.7   Benzoin condensation catalyzed by (a) cyanide; (b) thiamine.

3.1.6 Other Nucleophilic Additions

The condensation of olefins with aldehydes in the presence of an acid catalyst is referred to as the *Prins reaction* (36). Often, the reaction is carried out by using an aqueous solution of the aldehyde. For example, the reaction of 1-olefins with aqueous formaldehyde solution at elevated temperature gives 4-alkyl-1,3-dioxane and 3-alkyltetrahydropyran-4-ols (37):

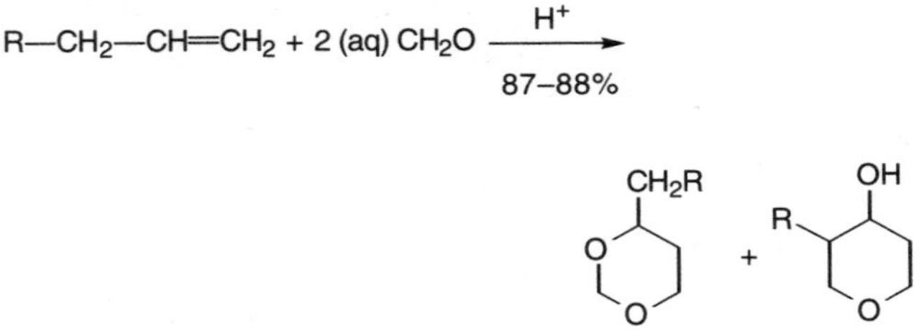

$$R-CH_2-CH{=}CH_2 + 2 \text{ (aq) } CH_2O \xrightarrow[87-88\%]{H^+}$$

(3.13)

Another related reaction is the Mannich-type reaction. The reaction is useful for the synthesis of β-aminoketones. The rate of Mannich reaction of phenols and ketones with secondary amines is far greater with aqueous solvents than with alcoholic or hydrocarbon solvents (38). Under an aqueous condition of THF–water (9:1), an aldehyde reacts with an amine and a vinyl ether to give the Mannich-type product, in the presence of 10 mol% of ytterbium triflate, Yb(OTf)$_3$ (39):

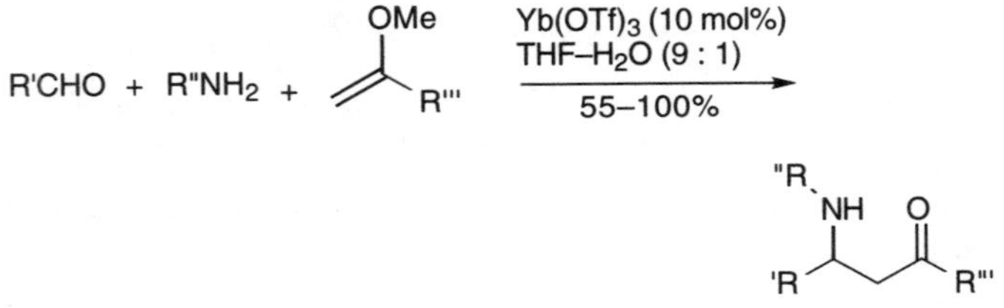

$$R'CHO + R''NH_2 + \underset{R'''}{\overset{OMe}{\diagup}} \xrightarrow[55-100\%]{\substack{Yb(OTf)_3 \text{ (10 mol\%)} \\ THF-H_2O \text{ (9 : 1)}}}$$

(3.14)

Wang recently reported a lanthanide-catalyzed reaction of indole with benzaldehyde [Eq. (3.15)]. The use of an ethanol/water system was found to be the best in terms of both yield and product isolation. The use of organic solvent such as chloroform resulted in oxidized byproducts (40).

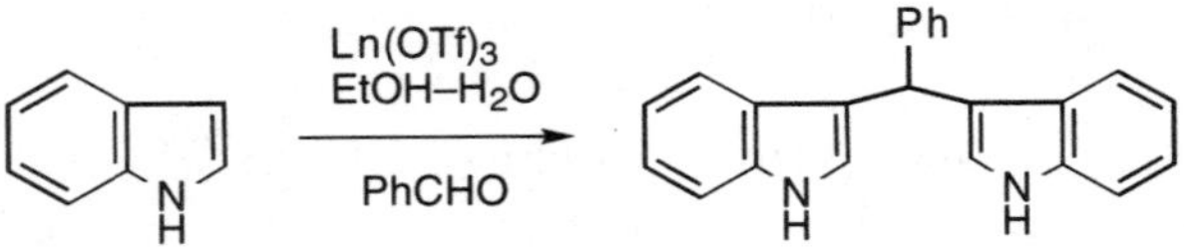

(3.15)

The reaction of aromatic radicals, generated by the decomposition of diazonium salts, with iminium salts in the presence of TiCl$_3$ in aqueous media produced secondary amines [Eq. (3.16)] (41). The iminium salts are formed *in situ* from aromatic amines and aldehydes.

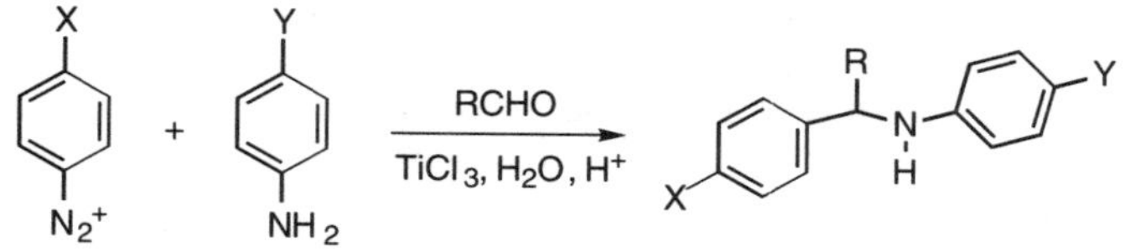

$$X = H, Cl, Me, OMe$$
$$Y = H, OMe$$
$$R = Me, Et, p\text{-}Meph$$

(3.16)

Petasis reported an efficient addition of vinyl boronic acid to iminium salts (42). Although no reaction was observed when acetonitrile was used as solvent, the reaction proceeded smoothly in water to give allyl amines [Eq. (3.17)]. The reaction of the boron reagent with

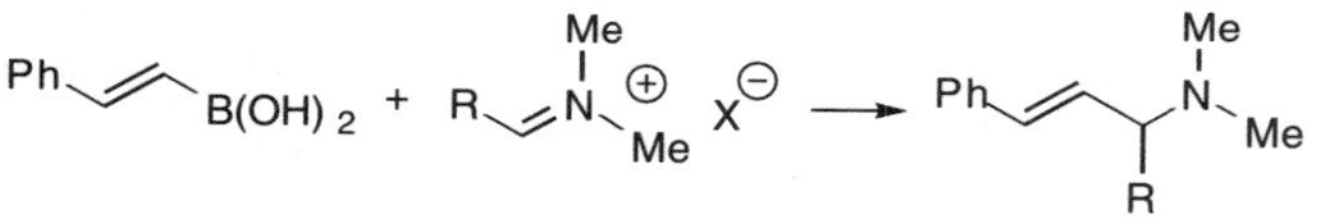

(3.17)

PhCH$_3$ or MeCN   No reaction

H$_2$O   85%

iminium ions, generated from glyoxylic acid and amines, affords novel α-amino acids [Eq. (3.18)].

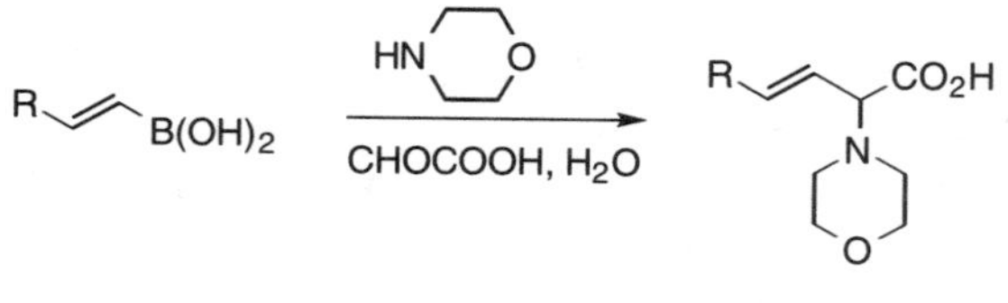

(3.18)

3.2 PHOTOCHEMICAL REACTIONS

Photoirradiation of dibenoyldiazomethane in the presence of aminoacid derivatives in aqueous solutions efficiently produced the addition products through a reaction with photogenerated carbene, which rearranged to the benzoylketene (Scheme 3.8) (43).

     The use of water as solvent interestingly influences the chemoselectivity in photochemical substitution reactions. For example, while the

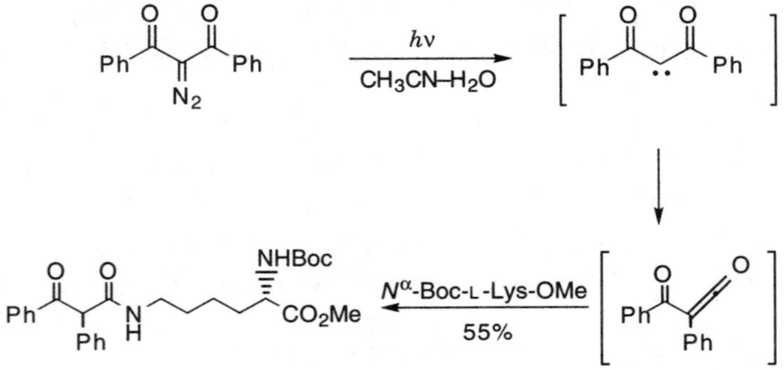

Scheme 3.8

photochemical aromatic substitution of fluorine by cyano group in *ortho*-fluoroanisole gives predominantly the hydroxylation product, the same reaction with *para*-fluoroanisole generates the cyanation product preferentially (Scheme 3.9) (44). The hydrogen bonding between water and the methoxyl group was attributed to the hydroxylation reaction in *ortho*-fluoroanisole. The effect of such a hydrogen bonding on the product distribution is much less in the latter case.

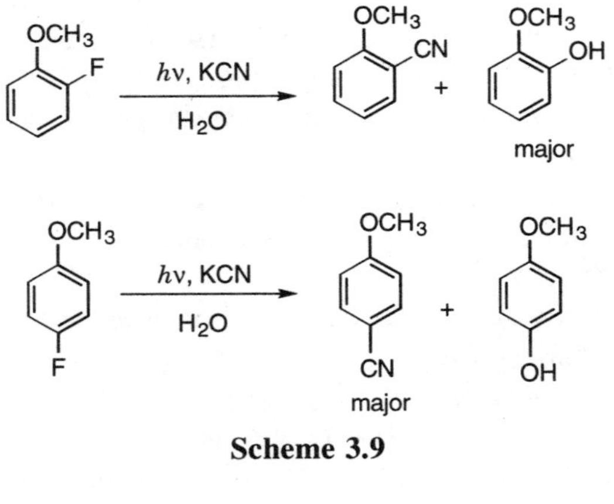

Scheme 3.9

REFERENCES

1. For a recent monograph on phase-transfer-catalysis, see Starks, C. M.; Liotta, C. L.; Halpern, M. *Phase-Transfer-Catalysis,* Chapman & Hall, London, 1994.

2. van Eldik, R.; Asano, T.; Le Nobel, W. J. *Chem. Rev.* **89**, 549 (1989).

3. Hajos, Z. G.; Parrish, D. R. *J. Org. Chem.* **39**, 1612 (1974).

4. Eder, U.; Sauer, G.; Wiechert, R. *Angew. Chem., Int. Ed. Engl.* **10**, 496 (1971).

5. Harada, N.; Sugioka, T.; Uda, H.; Kuriki, T. *Synthesis* 53 (1990).

6. Lavallee, J. F.; Deslongchamps, P. *Tetrahedron Lett.* **29**, 6033 (1988).

7. Keller, E.; Feringa, B. L. *Tetrahedron Lett.* **37**, 1870 (1996).

8. Lubineau, A.; Auge, J. *Tetrahedron Lett.* **33**, 8073 (1992).

9. Sussangkarn, K.; Fodor, G.; Karle, I.; George, C. *Tetrahedron* **44**, 7047 (1988).

10. Larpent, C.; Patin, H. *Tetrahedron* **44**, 6107 (1988).

11. Larpent, C.; Meignan, G.; Patin, H. *Tetrahedron* **46**, 6381 (1990); Larpent, C.; Meignan, G. *Tetrahedron Lett.* **34**, 4331 (1993).

12. Heo, C. K.; Bunting, J. W. *J. Org. Chem.* **57**, 3570 (1992).

13. For a recent review, see Basavaiah, D.; Rao, P. D.; Hymam, R. S. *Tetrahedron* **52**, 8001 (1996).

14. Augé, J.; Lubin, N.; Lubineau, A. *Tetrahedron Lett.* **35**, 7947 (1994).

15. For a review, see Nagata, W.; Yoshioka, M. *Org. React.* **25**, 255 (1968).

16. Lapworth, A.; Wechsler, E. *J. Chem. Soc.* **97**, 38 (1910).

17. For a recent example, see Ayed, T. B.; Amri, H. *Synth. Commun.* **25**, 3813 (1995).

18. For an excellent review, see Nielsen, A. T.; Houlihan, W. J. *Org. React.* **16**, 1 (1968).

19. Grignard, V.; Dubien, M. *Ann. Chim.* (Paris) **2**(10), 282 (1924).

20. Fringuelli, F. *Tetrahedron Lett.* **50**, 11499 (1994).

21. Lubineau, A. *J. Org. Chem.* **51**, 2142 (1986); Lubineau, A.; Meyer, E. *Tetrahedron* **44**, 6065 (1988).

22. Kobayashi, S.; Hachiya, I. *J. Org. Chem.* **59**, 3590 (1994); for a review on lanthanides catalyzed organic reactions in aqueous media, see Kobayashi, S. *Synlett.* 689 (1994).

23. Bounora, P. T.; Rosauer, K. G.; Dai, L. *Tetrahedron Lett.* **26**, 4009 (1995).

24. Ayed, T. B.; Amri, H. *Synth. Commun.* **25**, 3813 (1995).

25. Piechucki, C. *Synthesis* 187 (1976); Mikolajczyk, M.; Grzejszczak, S.; Midura, W.; Zatorski, A. *Synthesis*, 396 (1976).

26. Rambaud, M.; de Vecchio, A.; Villieras, J. *Synth. Commun.* **14**, 833 (1984).

27. Schimtt, M.; Bourguignon, J. J.; Wermuth, C. G. *Tetrahedron Lett.* **31**, 2145 (1990).

28. Friedrich, K., in Patai, S.; Rappoport, Z., *The Chemistry of Functional Groups,* Supplement C, Part 2, 1345, Wiley, New York, 1983.

29. Taillades, J.; Commeyras, A. *Tetrahedron* **30**, 2493 (1974).

30. Webber, J. M. in *Advances in Carbohydrate Chemistry*, Vol. 17, Wolfrom, M. L.; Tipson, R. S., eds., Academic Press, New York, 1962.

31. Fisher, E.; Passmore, F. *Ber.* **23**, 2226 (1890).

32. Kool, E. T.; Breslow, R. *J. Am. Chem. Soc.* **110**, 1596 (1988).

33. Breslow, R.; Conners, R. V. *J. Am. Chem. Soc.* **117**, 6601 (1995).

34. Breslow, R. D. *J. Am. Chem. Soc.* **80**, 3719 (1958).

35. Breslow, R. D.; Kool, E. *Tetrahedron Lett.* **29**, 1635 (1988).

36. Prins, H. J. *Chem. Weekbl.* **16**, 1072 (1919); for a review, see Adams, D. R.; Bhatnagar, S. P. *Synthesis* 661 (1977).

37. Stapp, P. R.; *J. Org. Chem.* **35**, 2419 (1970).

38. Tychopoulos, V.; Tyman, J. H. P. *Synth. Commun.* **16**, 1401 (1986).

39. Kobayashi, S.; Ishitani, H. *J. Chem. Soc. , Chem. Commun.* 1571 (1995).

40. Chen, D.; Yu, L.; Wang, P. G. *Tetrahedron Lett.* **37**, 4467 (1996).

41. Clerici, A.; Porta, O. *Tetrahedron Lett.* **31**, 2069 (1990).

42. Petasis, N. A. 212th ACS National Meeting, Orlando, 1996.

43. Nakatani, K.; Shirai, J.; Tamaki, R.; Saito, I. *Tetrahedron Lett.* **36**, 5363 (1995).

44. Liu, J. H.; Weiss, R. G. *J. Org. Chem.* **50**, 3655 (1985).

CHAPTER 4

METAL-MEDIATED REACTIONS

> Things will develop in the opposite direction when they become extreme.
>
> —Lao-tze (604–531 B.C.)

Since the discovery of the formation of alkylzinc compounds by Frankland from the reaction of iodoalkanes with metallic zinc (1), the use of organometallic reagents for organic synthetic purposes has been growing in scope and importance (2). The successes of Reformatsky (3), Barbier (4), and Grignard (5) are among the milestones in this development. Later progress includes the use of lithium and other metals. These organometallic reagents and their reactions are notorious for their sensitivity toward moisture and, when generated, often have to be handled in a drybox of inert environment, absolutely devoid of water. Thus, it seems surprising that one would consider carrying out these reactions in aqueous media. However, if one were to think of the very opposite, as observed by the ancient Chinese philosopher Lao-tze, then a reaction of this type, which is based on the premise of the exclusion of water, could conceivably be carried out in water if the proper conditions could be found. Such organometallic reactions, if successful, could well have profound implications in organic chemistry. One might speculate that if this (water-based organometallic reactions) is possible, then anything is possible.

In retrospect, one might have expected the development of organometallic reactions in aqueous media not to be such a conceptual hurdle.

As early as 1905 (6), it was known that the preparation of arylmercuric chlorides could be carried out in aqueous media. In the 1960s, tribenzylstannyl halide was produced in large scale in water (7). Wurtz-type reductive coupling of allyl halides proceeded in aqueous alcohol (8). These earlier reports would tend to indicate that some organometallic reagents are quite stable in water and may well exhibit interesting chemistry in water.

However, in spite of these early observations, the exploration of aqueous organometallic reactions did not start until more than two decades later. The vast majority of research has been carried out only within the last decade. As will be shown in this chapter, not only are such reactions possible, but these metal mediated reactions in water may well rival the classic Barbier–Grignard reactions in their synthetic usefulness. They also offer intriguing challenges in terms of understanding their mechanism.

4.1 BARBIER–GRIGNARD-TYPE REACTION

The Barbier–Grignard-type reaction is one of the most important reactions in organic chemistry. In their most general forms, the reagents used in this reaction serve as the nucleophiles leading to carbon–carbon bond formations. When the reaction electrophile is a carbonyl compound, the reactions are now frequently referred to as the *Barbier–Grignard-type reactions* [Eq. (4.1)]. The generation of the intermediate organometallic reagents can be in situ (Barbier) or stepwise (Grignard). A cardinal restriction of these reactions is the strict exclusion of moisture. Such a restriction undoubtedly imposes limitations in the application of these reactions in organic synthesis. In terms of synthetic design, various protecting group chemistries have to be implemented if acidic hydrogens are present in the substrates. In terms of actual laboratory practice, the solvent involved has to be dried thoroughly.

$$
\underset{R_1 \quad R_2}{\overset{O}{\parallel}} + \quad R'MX \quad \longrightarrow \quad \underset{R_1 \quad R_2}{\overset{OH}{\underset{}{\bigwedge}}}\!\!R' \qquad (4.1)
$$

In 1977, Wolinsky et al. (9) observed that the allylation reaction of carbonyl compounds with allyl bromide mediated by zinc could be carried out in 95% ethanol and *tert*-butanol. However, only a moderate yield was obtained. Even though this observation passed largely unnoticed, it signaled the possibility of carrying out organometallic reactions in hydroxylic solvents, including water. Since that time, significant progress has been made, and that same reaction can be carried out in aqueous medium through the use of a variety of metal mediators such as zinc, tin, indium, and bismuth. It is interesting to note that for the traditional Barbier–Grignard-type reactions in organic solvents, the metals used tend to be the more reactive ones such as magnesium or lithium, whereas for the aqueous reactions, the metals tend to be less metallic, or "softer."

The use of aqueous medium for Barbier–Grignard-type reactions offers considerable advantages (10). For instance, there is the practical convenience of not having to handle inflammable and anhydrous organic solvents. The tedious protection–deprotection processes for certain acidic-hydrogen-containing functional groups can be avoided, which contributes to an overall synthetic efficiency. Water-soluble compounds, such as carbohydrates, can be reacted directly without the need of derivatization; such processes also reduce the burden of solvent disposal and its impact on the environment.

4.1.1 Reactions Mediated by Tin

In 1983, Nokami et al. observed an acceleration of the reaction rate during the allylation of carbonyl compounds with diallyltin dibromide in ether by the addition of water to the reaction mixture (11). In one case, by using a 1:1 mixture of ether/water as solvent, benzaldehyde was alkylated in 75% yield in 1.5 h, while the same reaction gave only <50% yield in a variety of organic solvents, such as ether, benzene, or ethyl acetate, even after a reaction time of 10 h. The reaction was equally successful with a combination of allyl bromide, metal tin, and a catalytic amount of hydrobromic acid [Eqs. (4.2) and (4.3)] (Table 4.1). In the latter case, the addition of metallic aluminum powder or foil to

$$\text{(4.2)}$$

$$\text{(4.3)}$$

TABLE 4.1 Tin Mediated Allylation of Carbonyl Compounds

Aldehyde or Ketone	Allyl Halide	Yield (%)
PhCHO	Allyl bromide	73
Hexanal	Allyl bromide	70
MeCH=CHCHO	Allyl bromide	57
Citral	Allyl bromide	68
Cyclohexanone	Allyl bromide	76
2-Methylcyclohexanone	Allyl bromide	45
n-Hex-COMe	Allyl bromide	58
MeCOCH$_2$CH$_2$COOH	Allyl bromide	47
PhCOMe	Allyl bromide	50
PhCHO	1-Bromobut-2-ene	87
Acetaldehyde	1-Bromobut-2-ene	100
Hexaaldehyde	1-Bromobut-2-ene	97
Iso-butyraldehyde	1-Bromobut-2-ene	92
MeCH=CHCHO	1-Bromobut-2-ene	96
Citral	1-Bromobut-2-ene	81

the reaction mixture dramatically improved the yield of the product. The use of allyl chloride for such a reaction, however, was not successful. Reaction of crotyl bromide generated a mixture of diastereomers.

This reaction can also proceed intramolecularly. Using the combination of tin, aluminum, and hydrobromic acid in aqueous medium, ketones having allylic halide functionality such as **1** and **2** were cyclized to form 5- and 6-membered rings (Scheme 4.1) (12). Similar reactions occurred with aldehydes (13).

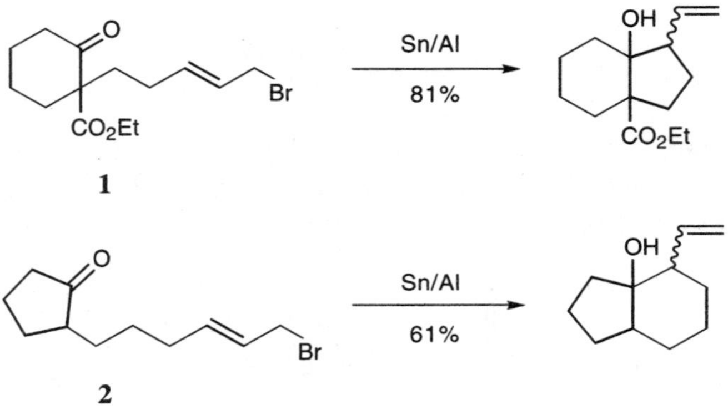

Scheme 4.1

Later, Torii et al. found that the tin–aluminum-mediated allylation can be carried out with less expensive allyl chloride, instead of allyl bromide, when a mixture of alcohol–water–acetic acid was used as the solvent (14). When combined with a stoichiometric amount of aluminum powder, both stoichometric and catalytic amounts of tin are effective. As reported by Wu et al., higher temperatures can be used to replace the use of aluminum powder (15). Under such a reaction condition, allyl quinones were obtained from 1,4-quinones followed by oxidation with ferric chloride [Eq. (4.4)]. Allylation reactions in water/organic solvent mixtures also utilized an electrochemical process, with the advantage that the allyltin reagent could be recycled (16).

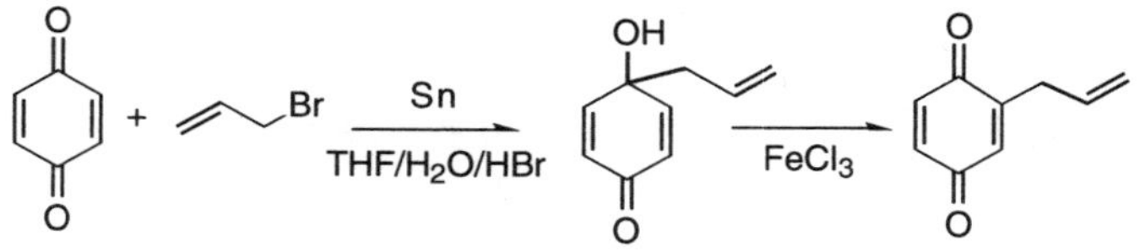

$$(4.4)$$

Otera et al. extended the tin mediated allylation reaction to 2-substituted allyl bromides (17). When 2-bromo- and 2-acetoxy-3-bromo-1-propene were used, the allylation reaction with tin produced the corresponding functionalized coupling products [Eq. (4.5)]. In the case of 2,3-dibromopropene, the reaction occurred exclusively through the allylic bromide in the presence of the vinyl group. The presence of other electrophiles, such as a nitrile (—CN) or an ester (—COOR), did not interfere with the reaction.

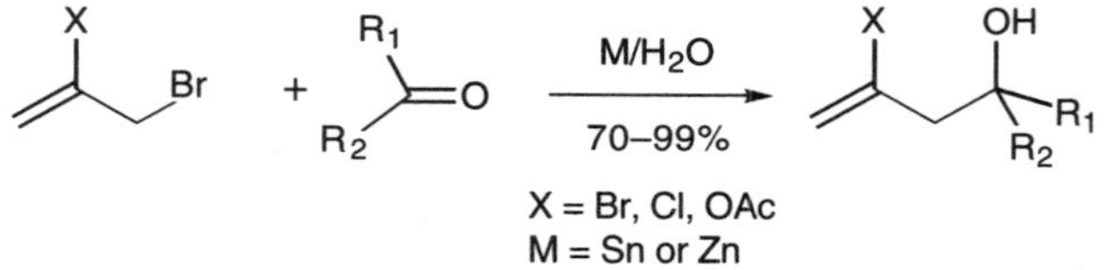

X = Br, Cl, OAc
M = Sn or Zn

$$(4.5)$$

Luche found that the tin-mediated allylation reactions can also be performed through ultrasonic irradiation, instead of using aluminum powder and hydrobromic acid to promote the reaction [Eq. (4.6)]

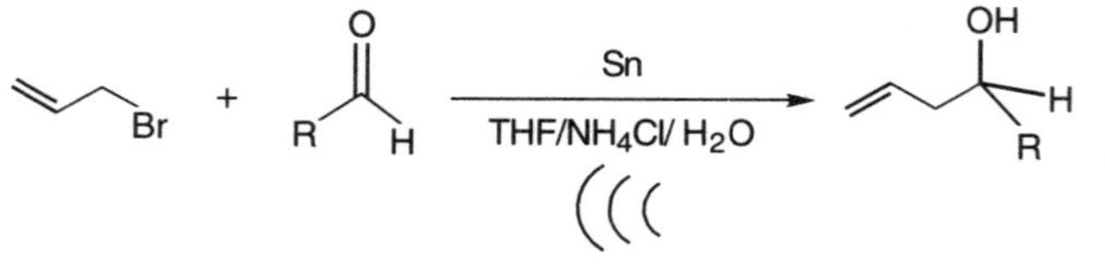

$$(4.6)$$

(18, 19). The use of saturated aqueous NH_4Cl/THF solution, instead of water/THF, dramatically increased the yield. When a mixture of aldehyde and ketone was subjected to the reaction, highly selective allylation of the aldehyde was achieved.

The allylation of carbonyl compounds in aqueous media with $SnCl_2$ can also employ allylic alcohols (20) or carboxylates (21) in the presence of a palladium catalyst [Eq. (4.7)]. The diastereoselectivity of the reactions with substituted crotyl alcohols was solvent-dependent. Improved diastereoselectivity was obtained with a mixture of water + THF or water + DMSO (dimethylsulfoxide) instead of using the organic solvent alone. For allyl bromides, $SnBr_2$ can be used in a mixture of methylene chloride/water (22).

$$\underset{CH_3}{\overset{R}{\diagup}}_{OH} \quad \xrightarrow[\text{2. PhCHO}]{\text{1. Pd(0), SnCl}_2} \quad \overset{R}{\diagup}\underset{Ph}{\overset{OH}{\diagdown}} \quad + \quad \overset{R}{\diagup}\underset{Ph}{\overset{OH}{\diagdown}} \quad (4.7)$$

The reaction of propargyl bromide with aldehydes mediated by tin in water generated a mixture of propargylation and allenylation products [Eq. (4.8)]. The selectivity in product formation is rather low (23).

$$\underset{R \quad H}{\overset{O}{\diagdown\!\!\!/}} + \equiv\!\!\diagup^{Br} \quad \xrightarrow{\text{Sn/H}_2\text{O/HBr}} \quad \underset{R}{\overset{OH}{\diagup}}\!\!\diagup\!\!= \quad + \quad \underset{R}{\overset{OH}{\diagup}}\!\!\diagup\!\!\equiv \quad (4.8)$$

Allylations, allenylations, and propargylations of carbonyl compounds in aqueous media also could be carried out with the preformed organic tin reagents, instead of using metals (24–27). As reported by Grieco et al. (28), allylation with preformed allylstannane can be applied to immonium salts generated *in situ* from primary amines and formaldehyde:

$$RNH_2 \cdot TFA + Bu_3SnCH_2CH{=}CH_2 \quad \xrightarrow[\text{H}_2\text{O–MeOH–CHCl}_3]{\text{HCHO}} \quad (4.9)$$

$$RNH(CH_2CH_2CH{=}CH_2)_2$$
$$(78{-}100\%)$$

The reactions of bromoacrylic acid and carbonyl compounds with Sn/Al (29), $SnCl_2/AcOH$ (30), or $SnCl_2/Amberlyst$ 15 (31) in aqueous media under refluxing conditions give 2-methylene-4-butyrolactones

[Eq. (4.10)] (**4**). A similar reaction occurs with the corresponding ester (**32**). Normal α-halo carboxylic esters do not react under these conditions. Since the substrate (**3**) is structurally an allylic halide, the reaction can be regarded as an allylation.

$$(4.10)$$

The aldol condensation is a related reaction for forming carbon–carbon bonds (33). However, under the classic aldol reaction conditions involving basic media, dimers, polymers, self-condensation products, or α,β-unsaturated carbonyl compounds often are formed as well. The formation of these products is attributed to the fact that the aldol condensation is an equilibrium process (34). Useful modifications of the classic aldol condensation, especially using Lewis acid–promoted reactions of enol silyl, or tin, ethers with carbonyl compounds (35, 36), have been developed to eliminate these difficulties. These modifications typically include the use of an organic solvent as the reaction media and require the exclusion of moisture.

Recently, the cross-aldol reaction of silyl enol ethers with carbonyl compounds was carried out in aqueous solvents without any acid catalyst. However, the reaction took several days to complete [Eq. (4.18)] (37). The cross-aldol products showed only a slight *syn* diastereoselectivity, which was the same as when this reaction was carried out in organic solvent under high pressure. Adding a catalytic amount of lanthanide triflate greatly improved the yields of such reactions (38). Among the lanthanide triflates, ytterbium triflate [Yb(OTf)$_3$], gadolinium triflate [Gd(OTf)$_3$] and lutetium triflate [Lu(OTf)$_3$] generally gave better yields of the aldol condensation product; the diastereoselectivities of these reactions were moderate. Water-soluble aldehydes were applicable, and the catalyst could be recovered and reused in this procedure.

The reaction of an α-halogen carbonyl compound with a metal and an aldehyde in water gave a direct cross-aldol reaction product through the Barbier-type reaction [Eq. (4.11)] (Table 4.2) (39).

$$(4.11)$$

TABLE 4.2 Tin-Mediated Direct Crossed Aldol Type Reaction

Halide	Aldehyde	*erythro/threo*	Yield (%)
PhCOCMe$_2$Br	PhCHO	—	41
PhCOCMe$_2$Br	n-C$_8$H$_{17}$CHO	—	57
PhCOCHMeBr	PhCHO	71:29	82
PhCOCHMeBr	c-C$_6$H$_{11}$CHO	64:36	83
PhCOCH$_2$Br	PhCHO	—	64
MeCOCHMe$_2$Br	PhCHO	—	85
MeCH$_2$COCH$_2$Br	PhCHO	—	17

A direct Reformatsky-type reaction occurred when an aromatic aldehyde reacted with an α-bromo ester in water mediated by tin [Eq. (4.12)] (40). However, the reaction gave only a low yield of the product, and aliphatic aldehydes failed to react under these conditions.

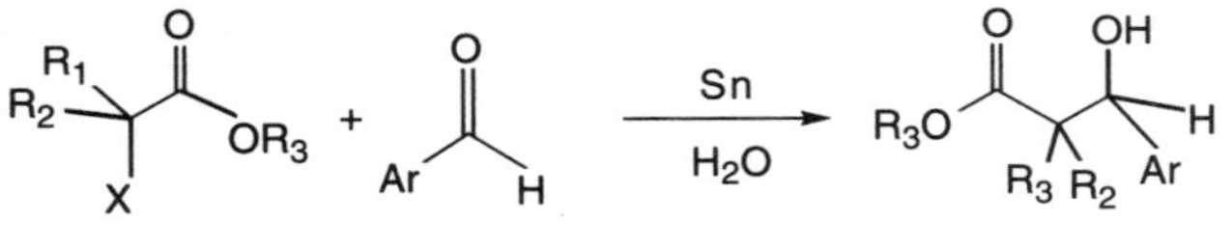

$$(4.12)$$

4.1.2 Reactions Mediated by Zinc

In 1977, Wolinsky et al. (41) reported that slow addition of allyl bromide to a stirred slurry of "activated" zinc dust and an aldehyde or a ketone in 95% ethanol or *tert*-butyl alcohol at a bath temperature of 78–95°C gave the allylation products in yields comparable to those obtained in aprotic solvents [Eq. (4.13)]. Prior to this report, Wurtz-type coupling of allyl halides was obtained in low yields in refluxing alcohol (42).

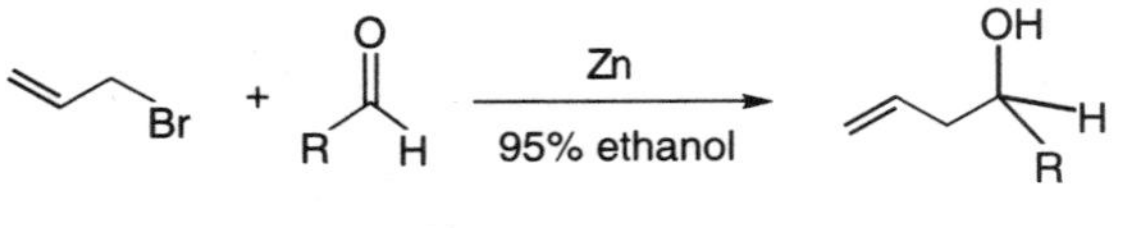

$$(4.13)$$

Then, in 1985, Luche et al. found that allylation of aldehydes and ketones can be effected in aqueous media using zinc as the metal and THF as a cosolvent under sonication conditions [Eq. (4.14)] (Table 4.3)

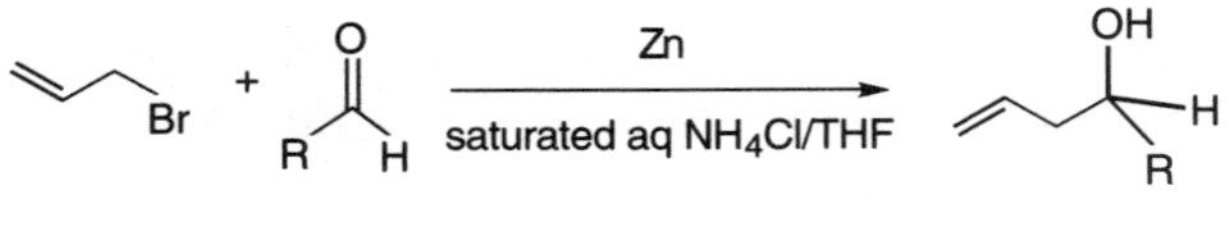

$$(4.14)$$

TABLE 4.3 Zinc-Mediated Allylation of Carbonyl Compounds

Susbtrate	Halide	Product	Yield (%)
PhCHO	Allyl bromide	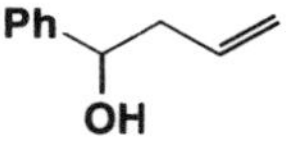	100
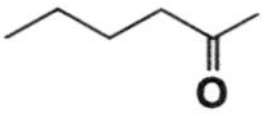	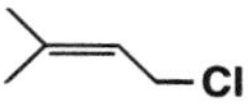	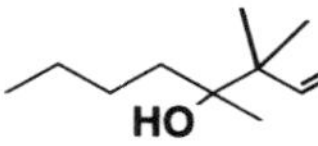	84
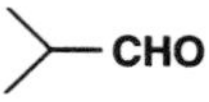	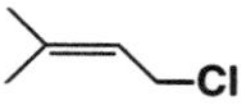	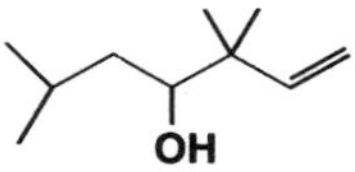	95
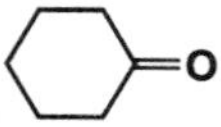	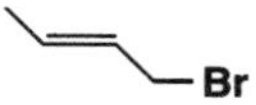	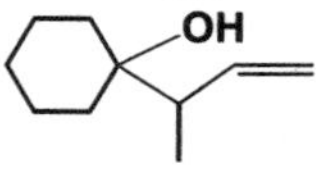	100
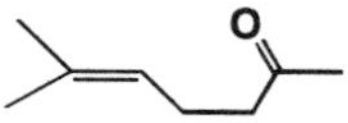	Allyl bromide	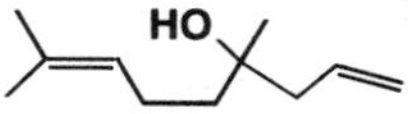	89
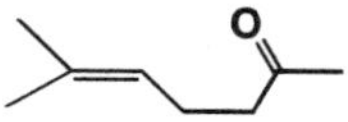	Allyl chloride	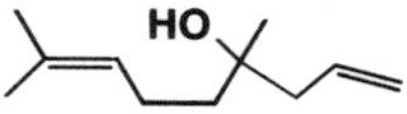	58

(43, 44). The replacement of water by saturated aqueous ammonium chloride solution improved the reaction significantly. Under such conditions, comparable results were obtained with or without the use of sonication.

In the same year, Benezra et al. reported (45) that ethyl (2-bromomethyl)acrylate can couple with carbonyl compounds, mediated by metallic zinc, in a mixture of saturated aqueous NH_4Cl-THF under refluxing conditions to give α-methylene-γ-butyrolactones [Eq. (4.15)]. The same reaction in THF alone gives only a low yield (15%) of the product. (2-Bromomethyl)acrylic acid could also be used directly on neutralization with triethylamine, but is much less effective.

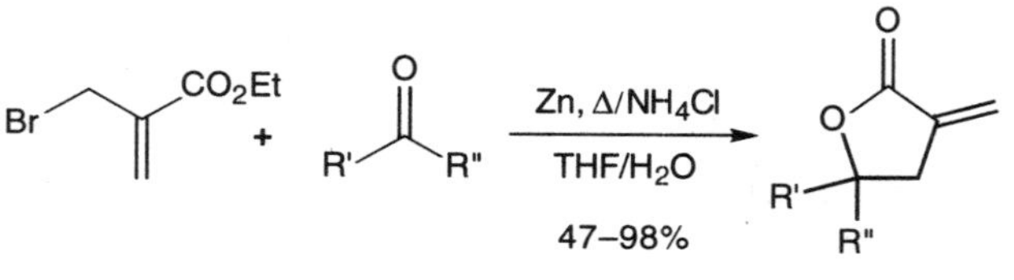

$$(4.15)$$

Later, Wilson carried out a more detailed study of zinc-mediated reactions in water (46) through a modification involving the use of a solid organic support instead of the cosolvent THF [Eq. (4.16)]. The solid or-

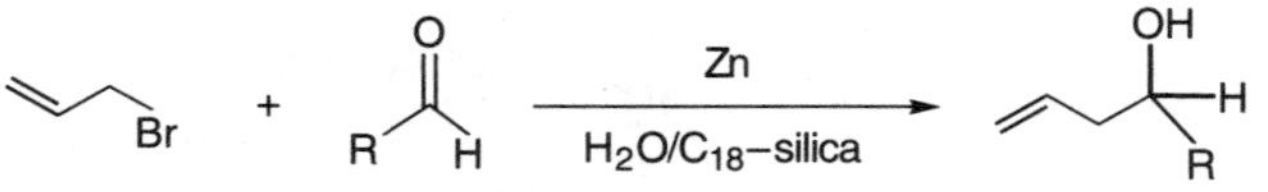

$$\text{(4.16)}$$

ganic supports include reverse-phase C_{18} silica gel, biobeads S-X8, a spherical porous styrene–divinylbenzene copolymer with 8% cross-links, and GC (gas chromatographic) column packing OV-101 on Chromosorb. The reactions proceed at about the same rate as reactions with THF as a cosolvent. Both allyl bromide and allyl chloride can be used.

Kunz and Reissig reported (47) the zinc-mediated allylation of methyl γ-oxocarboxylates (5) in a mixture of saturated aqueous ammonium chloride and THF [Eq. (4.17)]. The reaction provides a convenient synthesis of 5-allyl-substituted γ-lactone (6).

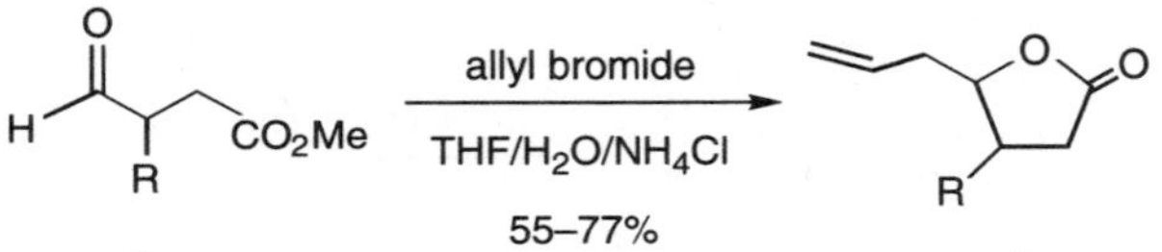

$$\text{(4.17)}$$

The reaction of an α-halogen carbonyl compound with zinc and an aldehyde in water gave a direct cross-aldol reaction product (39a). A direct Reformatsky-type reaction occurred when an aromatic aldehyde reacted with an α-bromo ester in water mediated by zinc [Eq. (4.18)] (48). Similar to the method using tin, the reaction gave only a modest yield of the product, and aliphatic aldehydes failed to react.

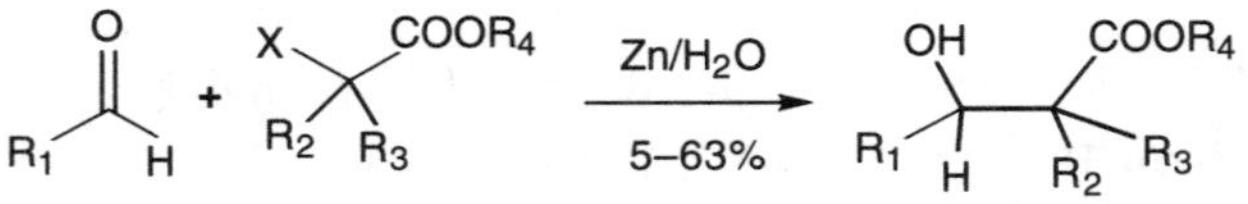

$$\text{(4.18)}$$

Chan and Li reported that conjugated 1,3-butadienes were produced in moderate yields when carbonyl compounds reacted with 1,3-dichloro-propene and zinc in water (49). The use of 3-iodo-1-chloropropene instead of 1,3-dichloropropene greatly improved the yields of the reactions. When the reactions were interrupted after their initial allylations, subsequent base treatment of the intermediate compounds produced vinyloxiranes in high yields (Scheme 4.2). Similarly, reactions of carbonyl compounds with 3-iodo-2-chloromethyl-1-propene followed by base treatment produced 2-methylenetetrahydrofurans

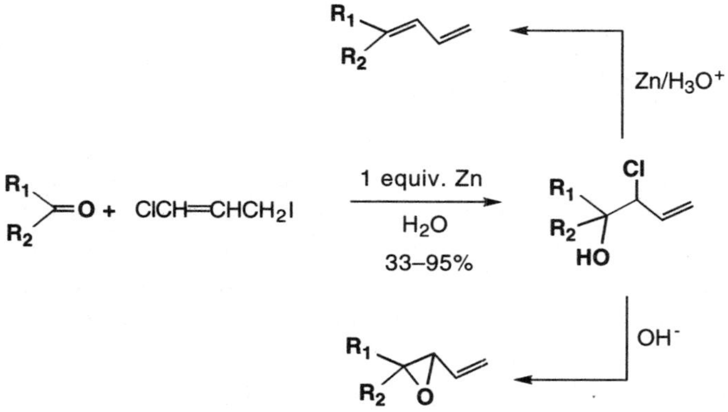

Scheme 4.2

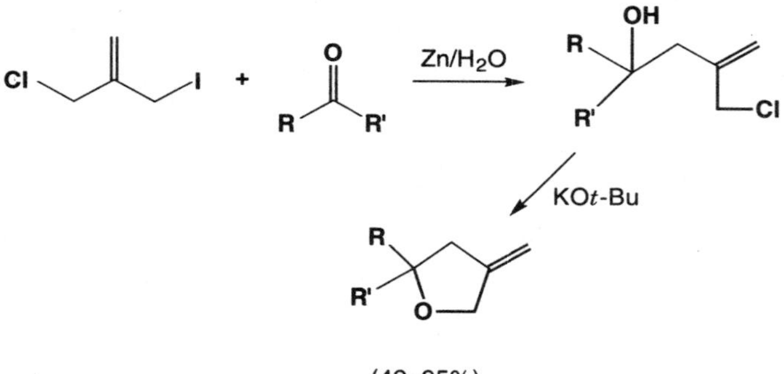

(42–95%)

Scheme 4.3

(Scheme 4.3) (50). Thus, the 3-iodo-2-chloromethyl-1-propene served as a novel trimethylenemethane equivalent (51).

Oda et al. reported that under refluxing conditions, the zinc-promoted reaction of 2,3-dichloro-1-propene with aldehydes and ketones in a two-phase system of water and toluene containing a small amount of acetic acid gave 2-chloroallylation products [Eq. (4.19)] (52). This reaction has been applied in a large-scale industrial synthesis (53).

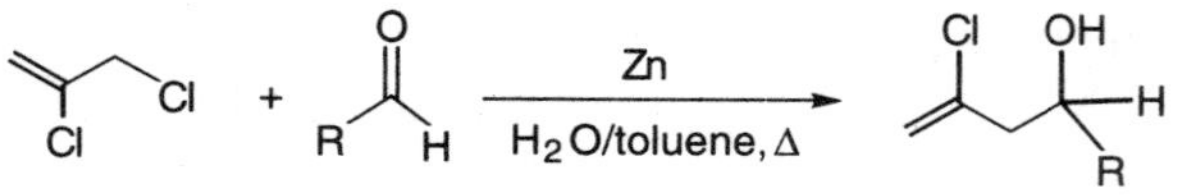

(4.19)

No reaction occurred when tin was used as the promoter. The absence of water completely shut down the reaction. It is interesting to note that the reaction of 2,3-dichloropropene with zinc powder in aqueous ethanol gives the dechlorination product, allene (54).

Very recently, Reisse used "activated" zinc for the aqueous Barbier-type reactions (55). Zinc powder produced by pulsed sonoelectroreduction is about three times more effective than the commercial one.

4.1.3 Reactions Mediated by Indium

In 1990, Li and Chan reported the first example of using indium to mediate Barbier–Grignard-type reactions in water (56). By examining the first ionization potentials of different elements, they pointed out that indium has a relatively low first ionization potential. In fact, the first ionization potential of indium is on the same level with the more reactive alkali metals and much lower than that of zinc or tin, or even magnesium (Table 4.4). On the other hand, the metal is not sensitive to boiling water or alkali and does not form oxides readily in air. Such special properties of indium indicate that indium can be a promising metal for aqueous Barbier–Grignard-type reactions. Indeed, it has since been found that indium is the most reactive and effective metal for such reactions.

TABLE 4.4 First Ionization Potential of Some Metals

Metal	Indium	Magnesium	Zinc	Tin	Lithium	Sodium
First	5.79	7.65	9.39	7.43	5.39	5.12
Ionization	—	—	—	—	—	—
Potential (eV)	—	—	—	—	—	—

Source: Obtained from *CRC Handbook of Chemistry and Physics*, 75th ed., CRC Press, Boca Raton, 1994.

When allylation reaction is mediated by indium in water, the reaction proceeds smoothly at room temperature without any promoter, while the use of zinc or tin usually requires acid catalysis, heat, or sonication (Table 4.5). An organic cosolvent is not necessary.

The mildness of the reaction conditions makes it possible to use the indium method to alkylate a methyl ketone in the presence of an acid-

TABLE 4.5 Indium-Mediated Allylation of Carbonyl Compounds

Substrate	Halide	Yield (%)
PhCHO	Allyl iodide	97
PhCHO	Allyl bromide	95
PhCHO	Allyl chloride	60
p-ClPhCHO	Allyl bromide	94
MeCH(OH)CHO	Allyl bromide	85
MeCH(ODCB)CHO	Allyl bromide	75
Me(CH$_2$)$_2$CH(OBn)CHO	Allyl bromide	90
PhCH(Me)CHO	Allyl bromide	72
Cyclohexanone	Allyl bromide	68
HOCH$_2$C(Me)$_2$CHO	Allyl bromide	85
HO(CH$_2$)$_2$CHO	Allyl bromide	95

sensitive acetal functional group [Eq. (4.20)]. The acetal function does not survive the reaction conditions with the methods of zinc or tin.

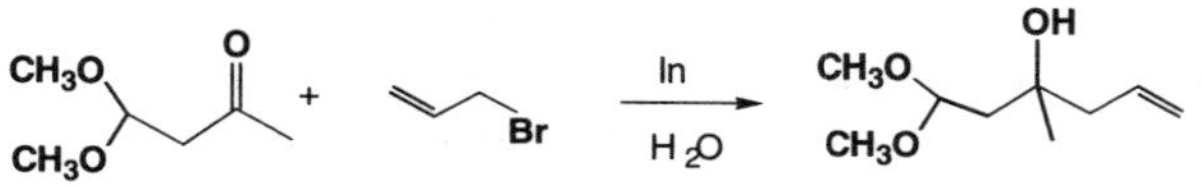

$\underline{M}$	Yield (%)
Zn	0 and destruction of starting materials
Sn	10 (under sonication)
In	70

(4.20)

The coupling of ethyl-2-(bromomethyl)acrylate with carbonyl compound proceeds equally well under the same reaction conditions. This provides a facile entry into the α-methylene-γ-lactone structure:

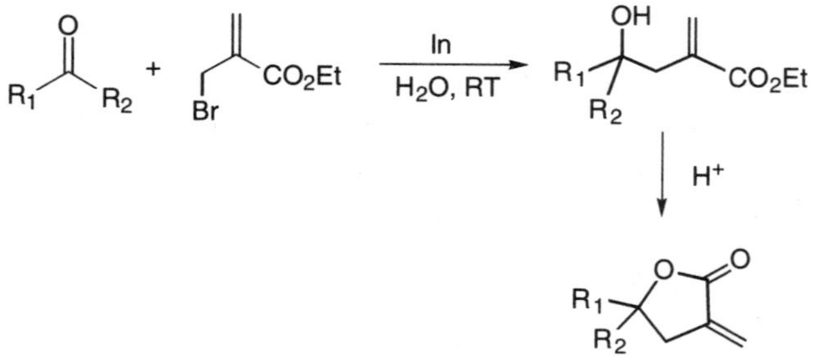

(4.21)

R$_1$ = Ph, R$_2$ = H   96%
R$_1$ = HOCH$_2$C(CH$_3$)$_2$, R$_2$ = H   85%

Later, it was found that the allylation of aldehydes and ketones can be carried out by using catalytic amounts of indium(III) chloride in combination with aluminum or zinc metal (57a). The reaction is typically performed in a THF–water (5:2) solvent mixture at room temperature. However, with this method, the reaction is much slower than the same reaction mediated by using stoichiometric amount of indium; it requires days to complete. When the reaction is carried out in anhydrous THF, the yield of the product drops considerably while the amounts of by-products, such as the reduction product alcohol, increase. The combinations of Al–InCl$_3$ or Zn–InCl$_3$ give comparable results.

Recently, Whitesides et al. (57b) examined the effect of substituents on the allyllic moiety on the indium-mediated reactions in water and found that reactions using indium at room temperature are comparable to tin-mediated reactions carried out at reflux. Replacement of the aqueous phase with 0.1 N HCl further increases the rate of the reaction. The reaction can also be carried out with preformed allylindium chloride.

More recently, Chan et al. reported that the carboxylic acid functionality on allyl halide is compatible with indium-mediated reactions [Eq. (4.22)] (48, 58). Thus, when 2-(bromomethyl)acrylic acid, instead of the ester, is reacted directly with carbonyl compounds and indium in water, the corresponding γ-hydroxyl-α-methylene carboxylic acids are generated in good yields. In some cases in which the indium powder and organic substrate clump together and make the stirring of the reaction mixture difficult, the addition of some ethanol is helpful. Interestingly, the addition of aprotic solvents, such as THF, completely shuts down the reaction.

$$\text{(4.22)}$$

Li reported that the reaction of 2-halomethyl-3-halo-1-propene with carbonyl compounds mediated by indium in water generates the bisallylation products [Eq. (4.23)] (59). The bromo compound is more effec-

$$\text{(4.23)}$$

7

tive for such reactions than the chloro one. Such a reaction effectively constitutes a trimethylenemethane dianion (**7**) equivalent in water.

An unusual regioselectivity was observed by Li et al. during the bisallylation reaction of 1,3-dibromo-propene with carbonyl compounds mediated by indium in water [Eq. (4.24)] (60). The reaction gives predominately 1,1-bisallylation product. Such a selectivity provides an allyl dianion (**8**) equivalent in water.

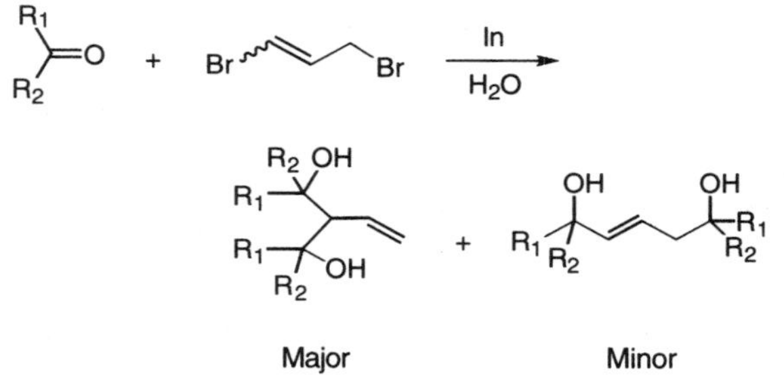

(4.24)

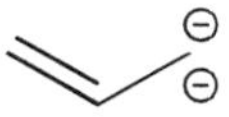

Major   Minor

8

Isaac and Chan (61) reported on the reaction of aldehydes with propargyl bromides in aqueous media mediated by indium [Eq. (4.25)] (Table 4.6). It has been found that the parent prop-2-yn-1-yl bromide reacts with both aliphatic and aromatic aldehydes in water to give mainly the homopropargyl alcohols. In contrast, when the propargyl bromide is γ-substituted, the coupling products are predominantly or exclusively the allenylic alcohols.

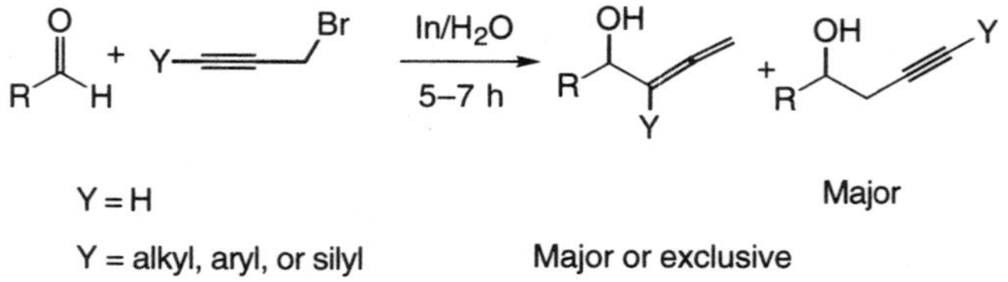

(4.25)

Y = H

Y = alkyl, aryl, or silyl   Major or exclusive   Major

Allylation of carbonyl compounds carried out with allylstannanes in combination with indium chloride in aqueous media has been reported by Marshall and Hinkle (62). Allylindium is proposed as the reaction intermediate.

TABLE 4.6 Indium-Mediated Coupling of Pro-2-ynyl Systems with Aldehydes

Aldehydes	Propargyl Bromide	Allene : Acetylene	Yield (%)
n-C_8H_{17}CHO	$BrCH_2CCH$	12 : 88	97
1-Naphthaldehyde	$BrCH_2CCH$	10 : 90	50
n-C_8H_{17}CHO	$BrCH_2CCPh$	95 : 5	89
HCOH	$BrCH_2CCPh$	99 : 1	94
$BuCCCH_2$CHO	$BrCH_2CCPh$	90 : 10	93
Cinnamaldehyde	$BrCH_2CCPh$	99 : 1	96
2-Furaldehyde	$BrCH_2CCPh$	93 : 7	75
n-C_8H_{17}CHO	$BrCH_2CCMe$	100 : 0	99
1-Naphthaldehyde	$BrCH_2CCMe$	100 : 0	<98
PhCHO	$BrCH_2CCSiMe_3$	80 : 20	60
n-C_8H_{17}CHO	$BrCH_2CCSiMe_3$	67 : 33	82
1-Naphthaldehyde	$BrCH_2CCSiMe_2Ph$	80 : 20	70
Me_2CHO	$BrCH_2CCSiMe_2Ph$	80 : 20	60

4.1.4 Reactions Mediated by Bismuth

In addition to the metals discussed above, other metals have been found to mediate Barbier–Grignard-type reactions in water, but investigation of these metals is somewhat limited. Wada reported (63) that metallic bismuth can also be used for the allylation reaction in aqueous media in a method similar to that used for tin, in which aluminum powder and hydrobromic acid are used as the promoter. Again, the reaction is more effective than the same reaction in an organic solvent. For purposes of comparison, when the allylation reaction of phenylacetaldehyde was carried out in a mixture of THF/water at room temperature, the corresponding alcohol was obtained in 90% yield. Under the same conditions, the use of THF alone as reaction solvent led to decreased yields and nonreproducible results. The uses of other metal promoters in combination with bismuth are also effective under the same reaction conditions. Such combinations include $Al(0)/BiCl_3$, $Zn(0)/BiCl_3$, and $Fe(0)/BiCl_3$.

Later, Katritzky et al. found that the bismuth(III)–aluminum system also mediated the allylation of immonium cations to give amines (Scheme 4.4) (Table 4.7) (64). In this case, even a methylation with iodomethane took place smoothly. Allylation of aldehydes carried out by electrochemically regenerated bismuth metal in an aqueous two-phase system has been reported by Tsuji (65).

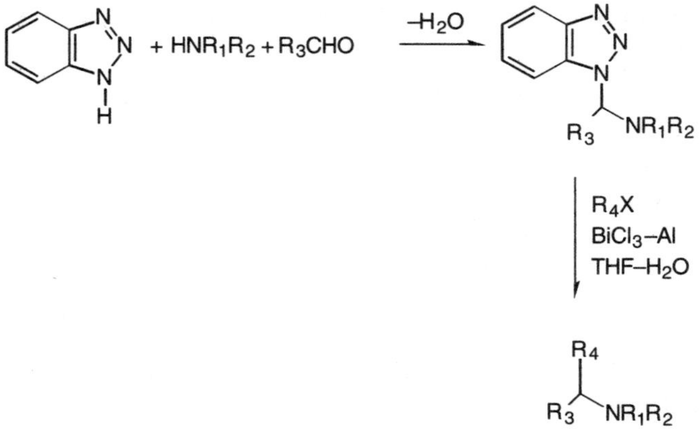

Scheme 4.4

TABLE 4.7 Amines Formation via Alkylation of Immonium Cations

R_1, R_2	R_3	R_4	Yield (%)
Ph, Me	H	Allyl	85
p-BuPh, Me	H	Allyl	87
Ph, Ph	H	Allyl	83
Bu, Bu	H	Allyl	83
i-Bu, i-Bu	H	Allyl	85
—(CH$_2$)$_2$O(CH$_2$)$_2$—	H	Allyl	80
—(CH$_2$)$_4$—	H	Allyl	85
—(CH$_2$)$_5$—	H	Allyl	87
c-C$_6$H$_{11}$, H	H	Allyl	80
p-MePh, H	H	Allyl	35
Ph(Me)CH, H	i-Pr	Allyl	80
MeCO, H	Ph	Allyl	78
PhCH$_2$OCO, H	Ph	Allyl	85
Ph, Ph	H	Propargyl	51
Ph,Ph	H	CH=C=CH$_2$	41
Ph, Me	H	Propargyl	48
Ph, Me	H	CH=C=CH$_2$	36
Ph, Me	H	Benzyl	75
Ph, Ph	H	Benzyl	79
—(CH$_2$)$_2$O(CH$_2$)$_2$—	H	Benzyl	70
t-Bu, i-Bu	H	Benzyl	88
Ph, Me	H	Ph(Me)CH	75
Ph, Ph	H	Me	75
i-Pr, i-Pr	H	Me	78
Ph, Me	c-C$_6$H$_{11}$	Me	60
Ph, Me	c-C$_6$H$_{11}$	H	30

4.1.5 Reactions with other Metals

Lead or cadmium powder can also be used for such reactions. These metals are much less reactive than those discussed previously (66). Several preformed organometallic reagents have been found to react with aldehydes in the presence of water. Kauffmann (67) found that while the chromium(III) compounds **9** are stable in aqueous solution at room temperature, they do not alkylate aldehydes. On the other hand, alkylchromium(III) reagents **10** and **11**, which are generated in THF, can react with aldehydes in the presence of water.

$$\{(H_2O)_5Cr-R\}^{2+} \qquad (THF)_3Cl_2Cr-R$$
$$\textbf{9} \;\; R = alkyl \qquad\qquad \textbf{10} \;\; R = alkyl$$

$$(THF)_3ClCr-R_2$$
$$\textbf{11} \;\; R = allyl \; or \; crotyl$$

The results indicated that, for the reaction of **10** (R = Me) with heptanal to give the addition product **12**, the yield of **12** increased initially with the addition of water, maximizing ($\leq 90\%$) at 1–2 mol of water per mole of Cr reagent [Eq. (4.26)]. Then the yield fell off as the amount of water increased further. However, in the reaction of benzaldehyde with the bisallyl reagent **11** (R = allyl), the yield of the product **13** remained high on further addition of water or ethanol [Eq. (4.27)].

$$n\text{-}C_6H_{13}CHO \;+\; \textbf{10} \xrightarrow[85\%]{} n\text{-}C_6H_{13}CH(OH)-CH_3 \qquad (4.26)$$
$$\textbf{12}$$

$$PhCHO \;+\; \textbf{11} \xrightarrow[52\%]{} PhCH(OH)-CH_2CH=CH_2 \qquad (4.27)$$
$$\textbf{13}$$

Kauffmann further reported that titanium(IV) reagents **14** and **15**, vanadium, niobium, and manganese(II) reagents **16–18** all alkylate aldehydes in ethanol. It is not clear whether similar reactions could also be carried out in water.

14 Cl_3Ti-Me
15 $(i\text{-}PrO)_3Ti\text{-}CH_2CH=CH_2$
16 $Cl-V(O)(CH_2CH=CH_2)_2$
17 $(EtO)_3NbMe_2$
18 $Mn(CH_2CH=CH_2)_2$

4.1.6 Chemoselectivity

Comparing Barbier–Grignard-type reactions in aqueous media and in organic solvents, one of the most striking differences is the high degree of chemoselectivity of the former. Usually, an aldehyde can be alkylated selectively in the presence of a ketone. As shown by Yamamoto et al. (68), the selectivity is usually higher than 99%. Even a cyclohexanone can be selectively allylated in the presence of cyclopentanone:

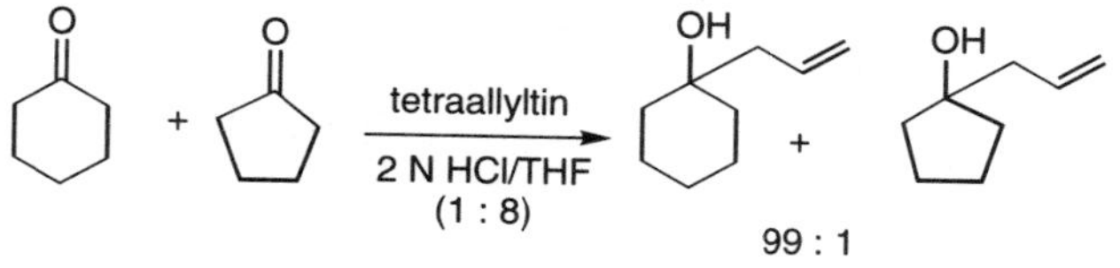

$$\text{(4.28)}$$

Other functional groups, such as esters, carboxylic acids, amides, phthalimides (69), nitriles, phosphonate esters (70), and acetals, are not reactive toward allyl bromide and indium in aqueous conditions. They can be present either in the substrates or as part of the allylic halide.

On the other hand, the nitro function appears to be reactive toward indium or tin in aqueous environments. In the attempted coupling of p-nitrobenzaldehyde with allyl iodide with indium in water, an orange insoluble polymeric product was obtained [Eq. (4.29)]. No allylation took place. It was assumed that the nitro function was reduced by indium to the corresponding amino group, which then polymerized with the aldehyde function.

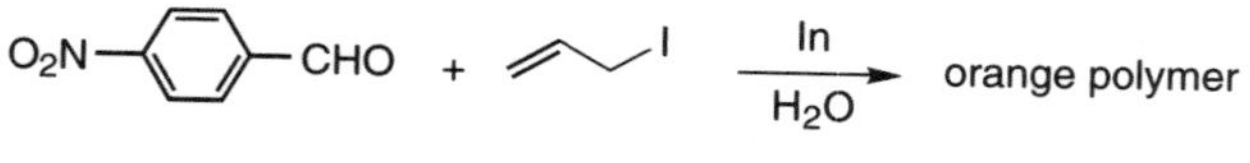

$$\text{(4.29)}$$

Bismuth, with a first ionization potential (7.29 eV) between that of indium and zinc, is, however, able to effect the coupling between p-nitrobenzaldehyde and allyl iodide in water to give the homoallylic alcohol in 70% isolated yield [Eq. (4.30)] (10b). In this reaction, the

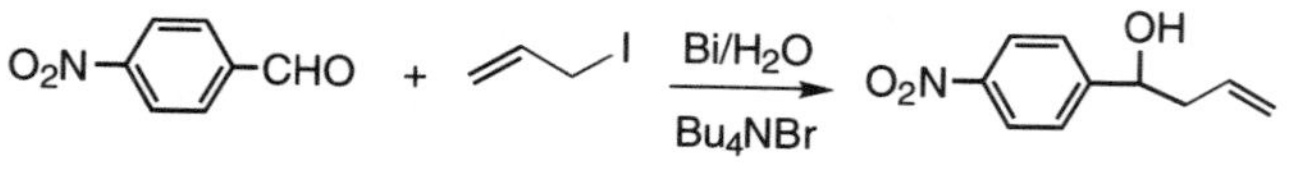

$$\text{(4.30)}$$

presence of tetrabutylammonium bromide facilitated the reaction considerably. The bismuth-mediated coupling is also quite sensitive to theallylic halide used. For example, allyl bromide or chloride are much less reactive under these conditions. These results suggest that the coupling reaction requires a finely tuned balance between the electron transfer from the metal to the different substrates in the reaction mixture.

Another example of the balance between substrates and metals is the rather different reaction pathways in the coupling of 1,3-dihalopropenes with carbonyl compounds, depending on the metal used. In the coupling of carbonyl compounds with 1-chloro-3-iodopropene with zinc in water, the chlorohydrin intermediates can be isolated, which, on further reaction with zinc, give the corresponding dienes [Eq. (4.31)] (49). On the other hand, in the coupling of carbonyl compounds with 1,3-dibromopropene and indium in water, the α,α-diadduct was obtained as the major product [Eq. (4.32)] (60). Presumably, in this case, the intermediate bromohydrin, which could be isolated, reacts with indium to give the diadduct at a rate competitive with elimination to the diene.

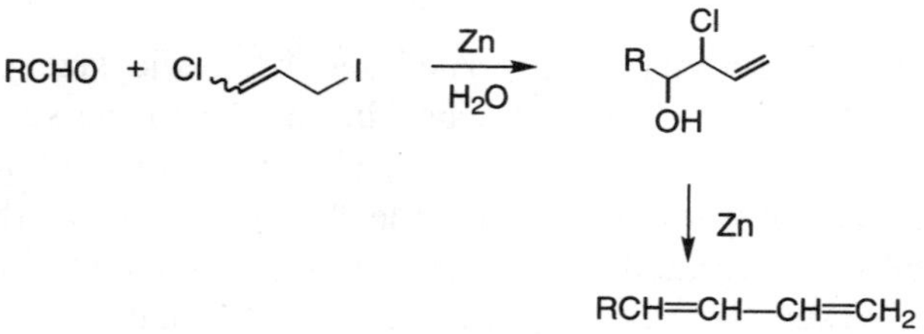

$$RCHO + Cl\diagup\diagdown I \xrightarrow[H_2O]{Zn} R\diagup\diagdown \overset{Cl}{\diagup}\diagdown$$

(4.31)

$$\downarrow Zn$$

$$RCH{=}CH{-}CH{=}CH_2$$

$$RCHO + Br\diagup\diagdown Br \xrightarrow[H_2O]{In} R\diagup\diagdown \overset{Br}{\diagup}\diagdown$$

$$\downarrow \begin{array}{l} In/H_2O \\ RCHO \end{array}$$

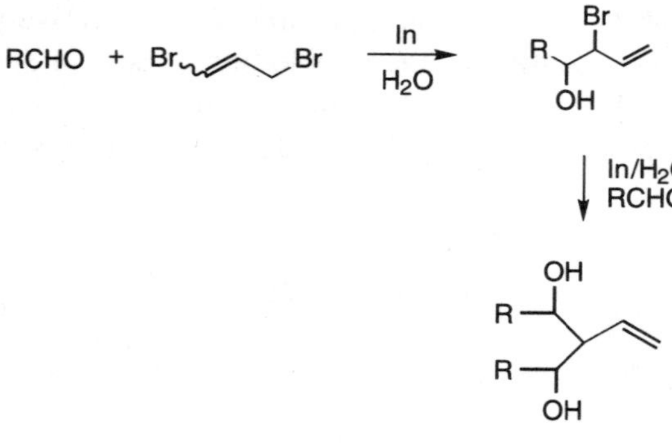

(4.32)

4.1.7 Regioselectivity

In the metal-mediated coupling of carbonyl compounds with substituted allylic halides in water, regioisomers **19** and **20** can be formed:

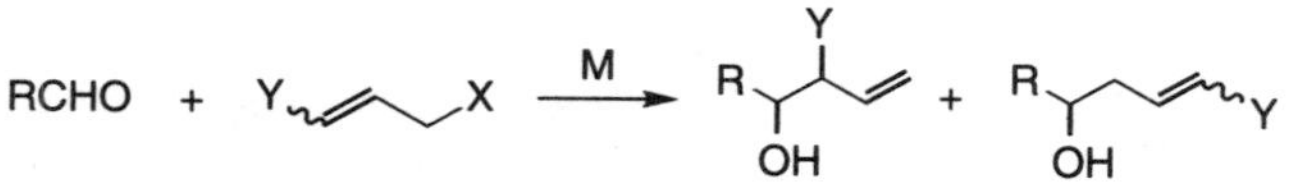

$$RCHO + Y\diagup\diagdown X \xrightarrow{M} R\diagup\diagdown + R\diagup\diagdown Y$$

(4.33)

19   **20**

With tin or zinc as the metal, there has been no systematic study on the regioselectivity of the reaction. Recently, the regioselectivity of indium-mediated coupling has been examined (71). The following conclusions can be drawn:

1. In general, the reaction gives the regioisomer where the substituent is alpha to the carbon–carbon bond to be formed. Thus, in the coupling of crotyl bromide with benzaldehyde by indium in water, the product is exclusively the γ-coupled isomer **21**:

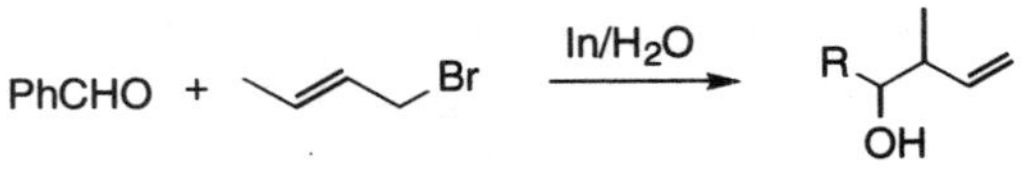

$$\text{(4.34)}$$

21

2. The regioselectivity is governed by the steric size of the substituent but not by the degree of substitution. When the susbstituent is sterically bulky (e.g., *tert*-butyl or silyl), the preferred regioisomer formed has the substituent away (at the γ-position) from the carbon–carbon bond being created. This is illustrated by the coupling of isobutanal with either the allyl bromide **22** or **23**. On the other hand, γ,γ-dimethylallyl bromide or the pinenyl bromide (**24**) reacted with benzaldehyde to give the adducts **25** and **26** respectively, despite the high degree of substitution at the double bond [Eqs. (4.35)–(4.38)]. Interestingly, the allyl bromide **27**, which had an essentially neopentyl type of substituent, appeared to behave as a sterically bulky substituent.

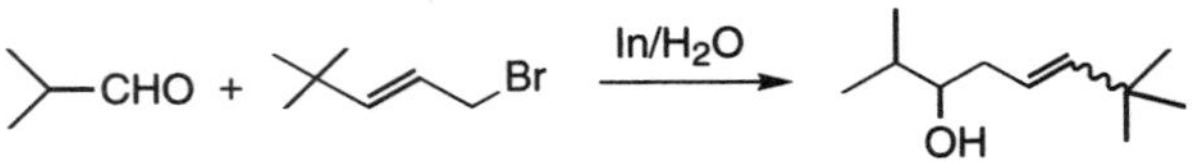

$$\text{(4.35)}$$

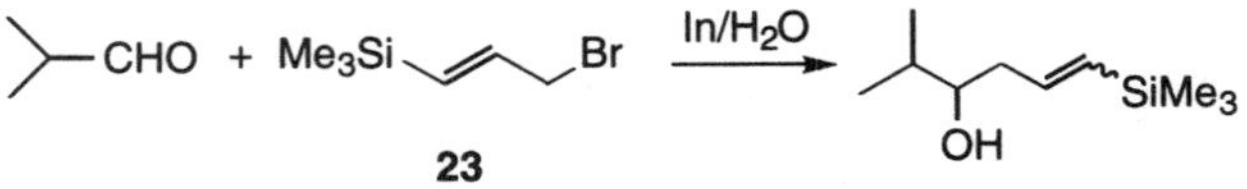

22

$$\text{(4.36)}$$

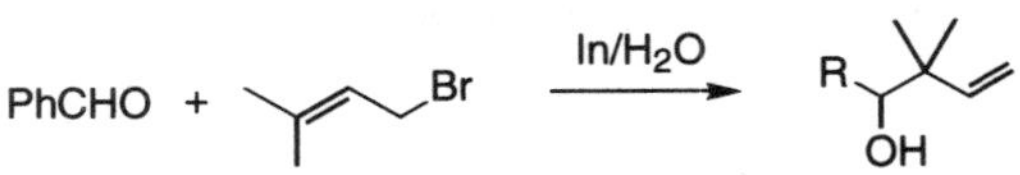

23

$$\text{(4.37)}$$

25

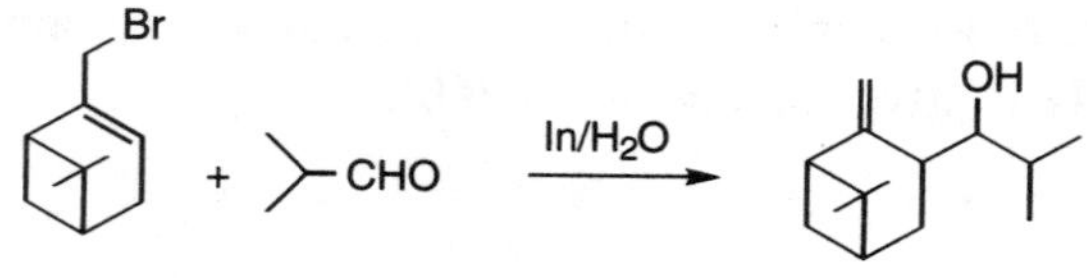

(4.38)

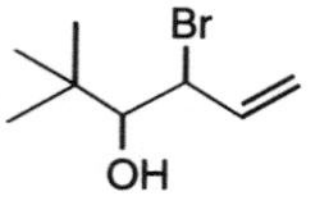

27

3. Regioselectivity is not governed by the conjugation of the double bond with the substituent. This is evident in the coupling of *E*-cinnamyl bromide or methyl 4-bromo-*E*-crotonate with isobutanal [Eqs. (4.39) and (4.40)]. In both cases, the deconjugated products **28** and **29** were obtained.

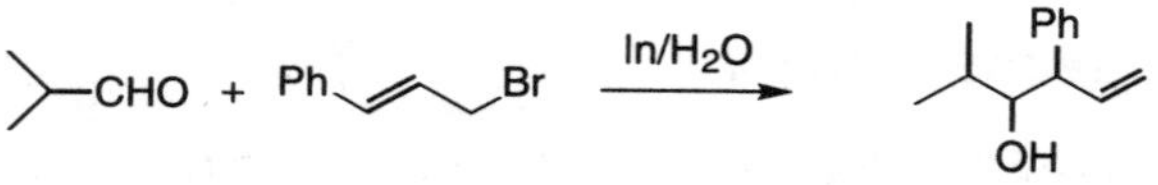

(4.39)

28

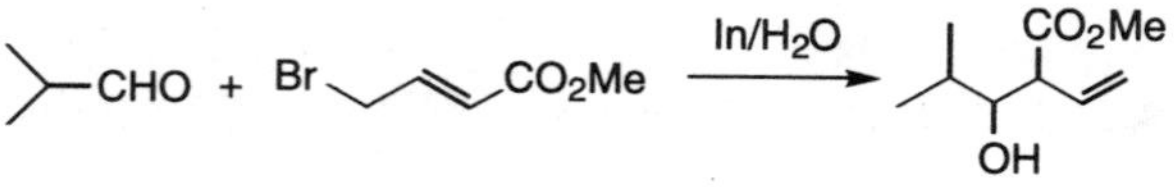

(4.40)

29

4. Regioselectivity is independent of the geometry of the double bond. Either *E*- or *Z*-cinnamyl bromide coupled with isobutanal to give the same regioisomer **28**. The possibility that the *Z*-cinnamyl bromide may have isomerized to the *E*-cinnamyl bromide first before coupling was ruled out by the observation that the *Z*-cinnamyl bromide could be recovered unchanged under the reaction conditions when the reaction was half-completed.

5. Regioselectivity is independent of the initial location of the substituent on the double bond. In the reaction of 1,3-dibromopropene with benzaldehyde, the intermediate allylic bromide **30** must have the bro-

mine α- to the substituent, and the second coupling must have occurred regioselectively to give the adduct **31** (60):

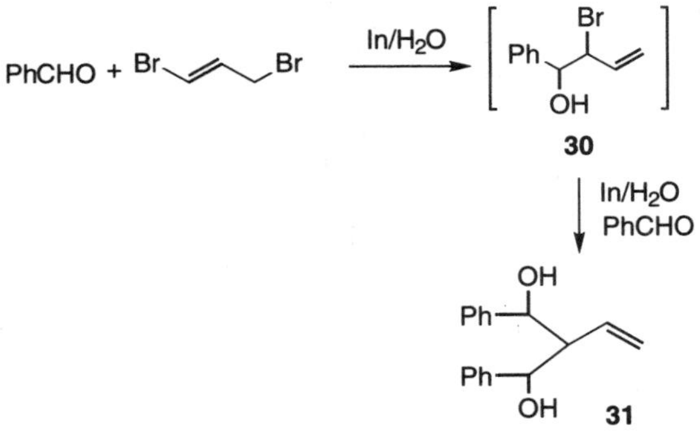

(4.41)

The regioselectivity in the coupling of aldehydes with substituted propargyl halides mediated by indium in water was discussed previously [Eq. (4.25)].

4.1.8 Diastereoselectivity

In terms of diastereoselectivity, there are two possible situations (types A and B in Scheme 4.5).

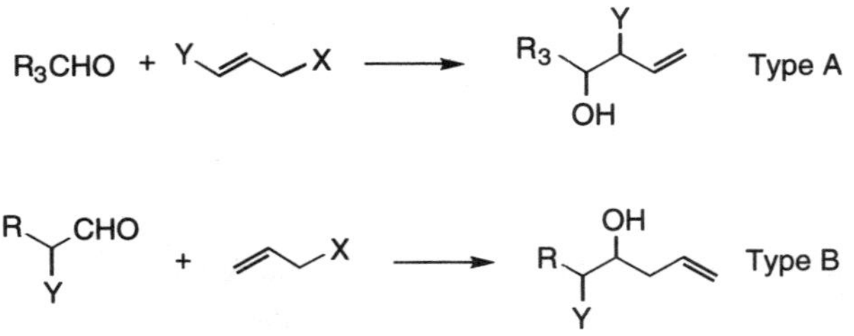

Scheme 4.5

Within the type A situation, most of the studies have concentrated on the indium-mediated reactions. The diastereoselectivity depends on the substituents on both the aldehyde and the allylic halide, but not on the geometry of the double bond of the allylic halide. This can be illustrated by the allylation of benzaldehyde with the three different substituted allylic halide as **32**, **33**, and **34** [Eqs. (4.42)–(4.44)] (71). The

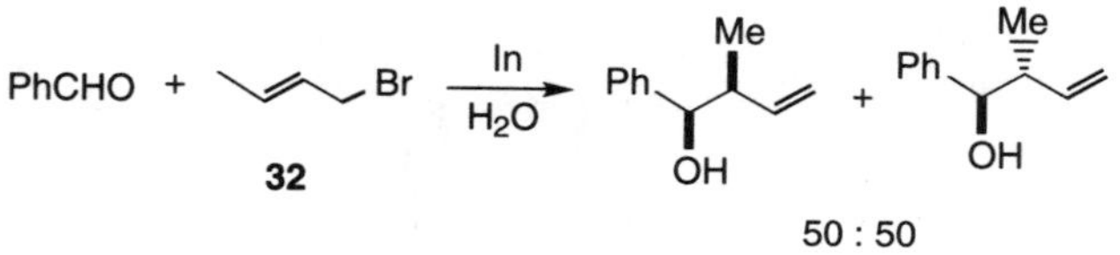

PhCHO + **32** (allyl bromide) $\xrightarrow[H_2O]{In}$ Ph—CH(OH)—CH(Me)—CH=CH₂ + Ph—CH(OH)—CH(Me)—CH=CH₂   (4.42)

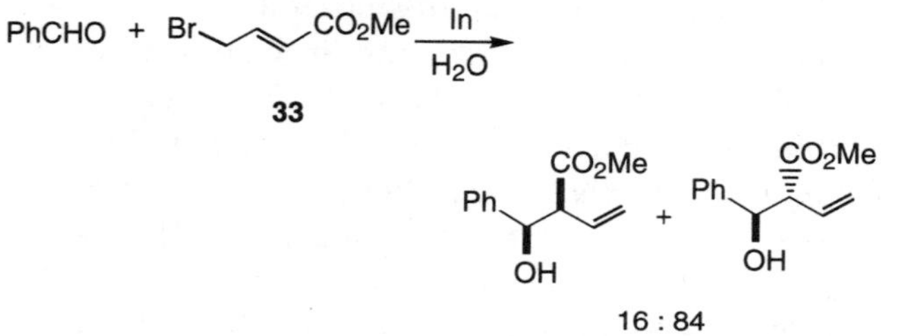

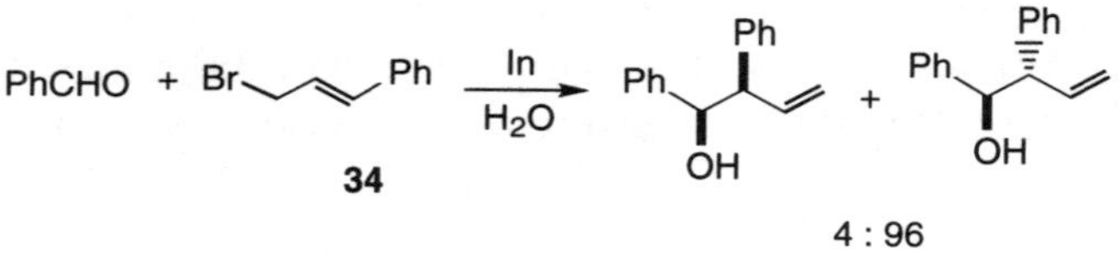

50 : 50

PhCHO + Br—CH=CH—CO₂Me **33** $\xrightarrow[H_2O]{In}$

Ph—CH(OH)—CH(CO₂Me)—CH=CH₂ + Ph—CH(OH)—CH(CO₂Me)—CH=CH₂   (4.43)

16 : 84

PhCHO + Br—CH=CH—Ph **34** $\xrightarrow[H_2O]{In}$ Ph—CH(OH)—CH(Ph)—CH=CH₂ + Ph—CH(OH)—CH(Ph)—CH=CH₂   (4.44)

4 : 96

selectivity can range from 1:1 in the case of **32** to an *anti* preference as high as 96:4 (*anti*:*syn*) in the case of **34**. For **34**, either the *E*- or the *Z*-cinnamyl bromide can be used, and nearly the same diastereoselectivity can be obtained. On the other hand, using the same substituted allylic halide, such as **34** with different aldehydes, the degree of diastereoselectivity seems to be dependent on the steric size of the substituent on the aldehyde. As the size increases from **35** to **36** to **37**, the *anti* diastereoselectivity increases as well [Eqs. (4.45)–(4.47)].

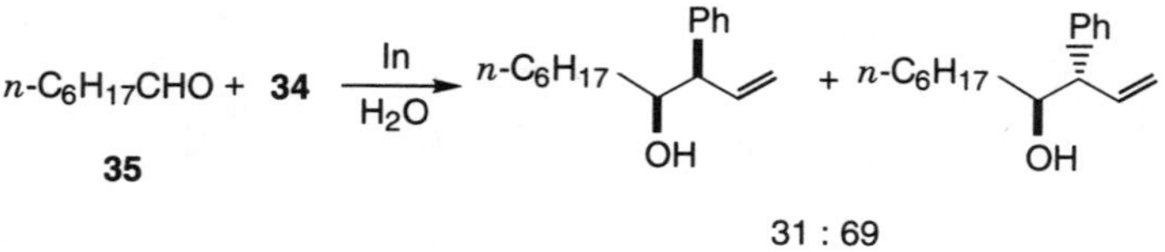

n-C₆H₁₇CHO + **34** $\xrightarrow[H_2O]{In}$ n-C₆H₁₇—CH(OH)—CH(Ph)—CH=CH₂ + n-C₆H₁₇—CH(OH)—CH(Ph)—CH=CH₂   (4.45)

35

31 : 69

C₆H₁₁—CHO + **34** $\xrightarrow[H_2O]{In}$ C₆H₁₁—CH(OH)—CH(Ph)—CH=CH₂ + C₆H₁₁—CH(OH)—CH(Ph)—CH=CH₂   (4.46)

36

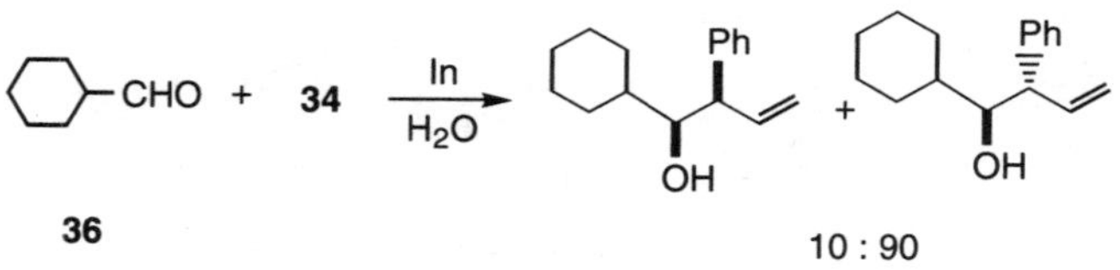

10 : 90

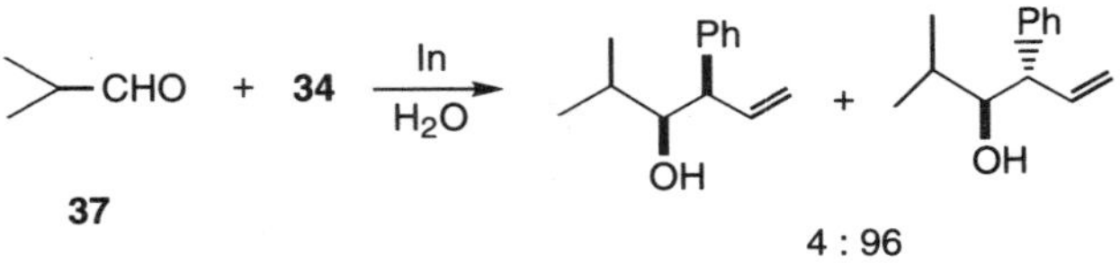

$$4 : 96 \tag{4.47}$$

To account for the regioselectivity observed above, as well as the *anti* diastereoselectivity, the following mechanism for the indium-mediated coupling of allylic bromides with aldehydes in aqueous media has been proposed (Scheme 4.6). First, an allyl indium species **38** is formed. Compound **38** exists in equilibrium with its stereoisomer **39** as well as its regioisomer **40**. In the coupling with aldehyde, the reaction proceeds through several possible cyclic transition states, **41** and **42**, all with the carbonyl oxygen coordinated with indium. In cases where the substituent (R_1 or R_2) is sterically bulky (e.g., silyl or *tert*-butyl), the cyclic transition state **42a** or **42b** is preferred, giving a mixture of *E*- and *Z*-substituted alkenes. For the other allylic bromides, the cyclic transition state **41a** or **41b** is favored. The diastereoselectivity is then

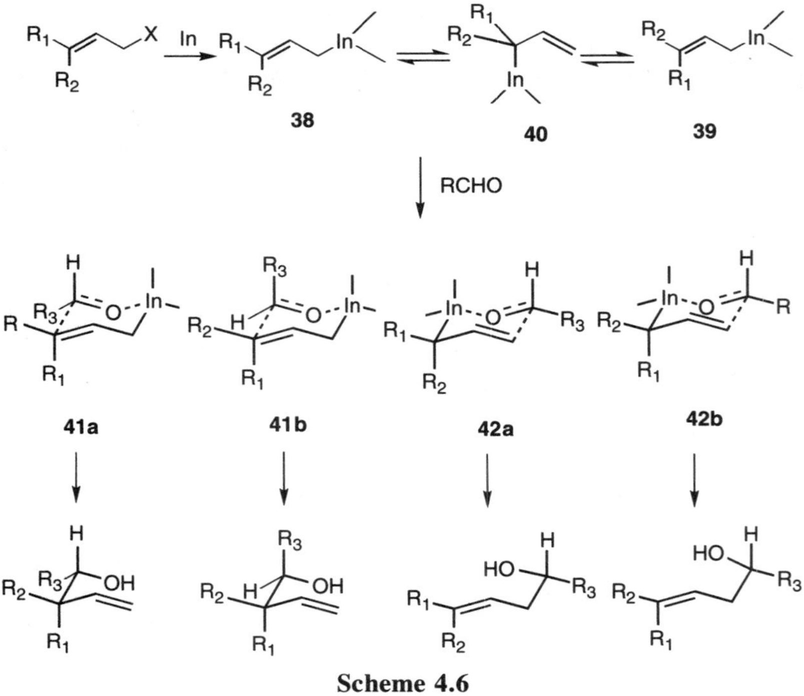

Scheme 4.6

governed by the steric size of the substituent on aldehyde (R_3) in differentiating between **41a** and **41b**. In general, the *anti* product would be favored.

Similar diastereoselectivity is observed in the reaction of aldehydes with cinnamyl chloride mediated by tin chloride/Al(0) [Eq. (4.48)] (72). In such a reaction, the diastereoselectivity is higher than 90:10 favoring the *anti*-isomer.

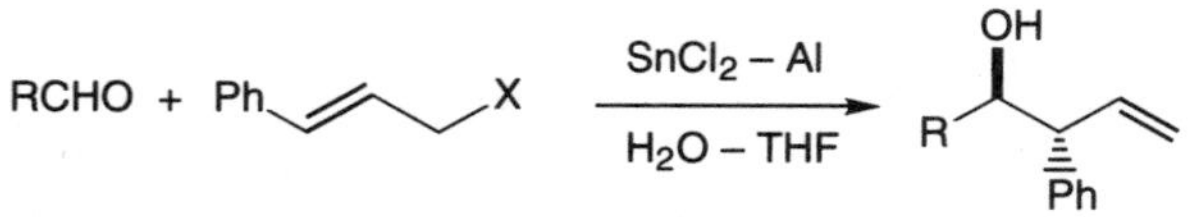

$$RCHO + Ph\diagdown\diagup X \quad \xrightarrow[\text{H}_2\text{O} - \text{THF}]{\text{SnCl}_2 - \text{Al}} \quad R\overset{\text{OH}}{\diagup}\diagdown\diagup \quad (4.48)$$
$$\underset{\text{Ph}}{}$$

In the type B situation, the reaction can favor either *syn* or *anti* diastereoselectivity, depending on the properties of the α-substituents. Using compound **43** as the model aldehyde with an adjacent chiral center, the coupling of **43** with allyl bromide/zinc in water showed a *syn* diastereoselectivity, the same selectivity as the reaction of **43** with other allylorganometallic reagents in organic solvents [Eq. (4.49)] (73). The *syn* selectivity can be accounted for using Cram's rule or its variants.

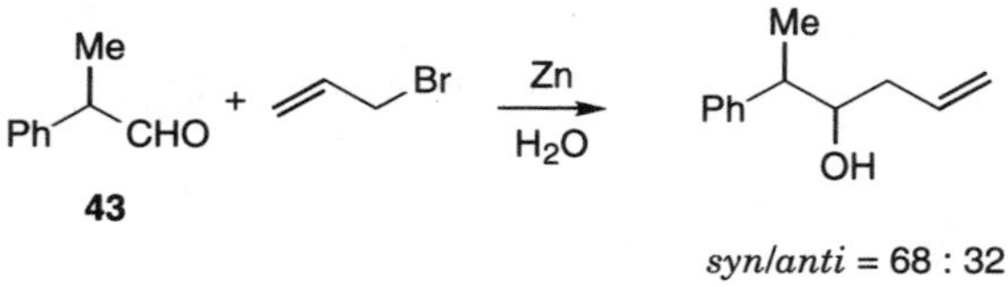

$$\underset{\textbf{43}}{\overset{\text{Me}}{\underset{\text{Ph}}{\diagup}}\text{CHO}} + \diagup\diagdown\text{Br} \quad \xrightarrow[\text{H}_2\text{O}]{\text{Zn}} \quad \overset{\text{Me}}{\underset{\text{OH}}{\text{Ph}}}\diagdown\diagup\diagdown\diagup \quad (4.49)$$

$$syn/anti = 68 : 32$$

Normally, the addition of C-nucleophiles to chiral α-alkoxyaldehydes in organic solvents is opposite to Cram's rule. The anti-Cram selectivity has been rationalized on the basis of chelation control as illustrated in Scheme 4.7 (74). Recently, the allylation of several α-

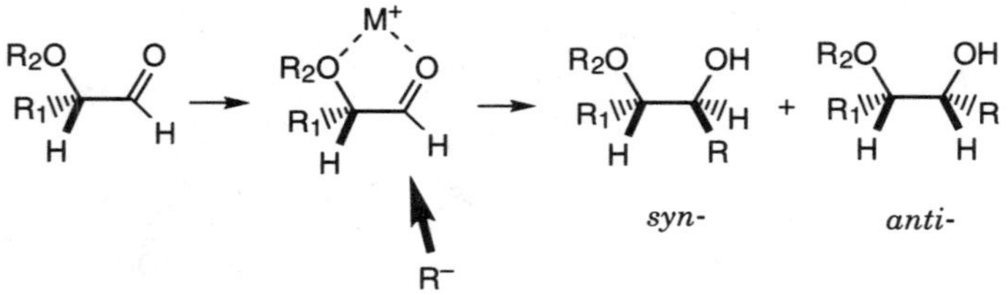

Scheme 4.7

alkoxyaldehydes with allylmagnesium bromide in ether is compared with the reaction with allyl bromide/zinc in water [Eq. (4.50)] (72). In the cases of reactions in organic solvent, the anti-Cram *syn*-isomer is obtained as expected from the chelation model. However, in the cases of reactions in water, a reversal of stereoselectivity is observed, with the *anti*-isomer as the major product. It is argued that in the aqueous reactions, the chelation model no longer operates for α-alkoxyaldehydes.

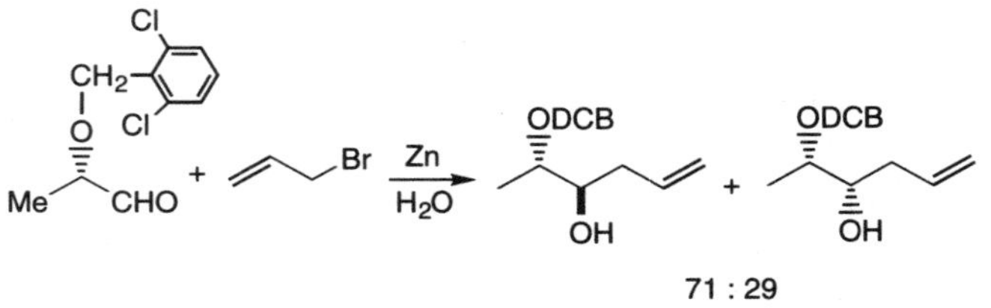

$$71:29 \tag{4.50}$$

The same *anti* preference was observed in the reactions of α-alkoxyaldehydes with allyl bromide/indium in water [Eq. (4.51)] (75).

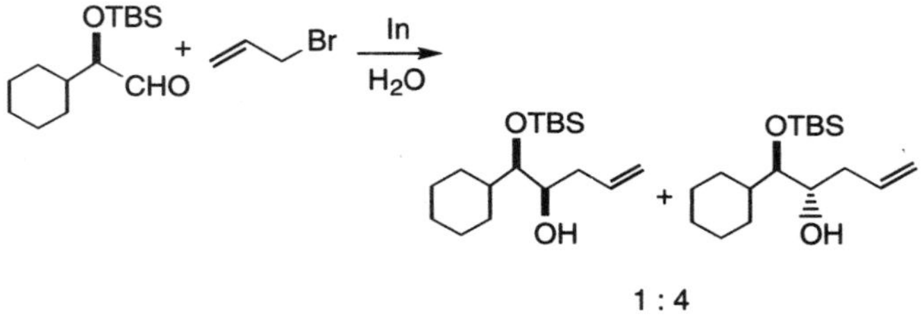

$$1:4 \tag{4.51}$$

However, for the allylation of α-hydroxyaldehydes with allyl bromide/ indium, the *syn*-isomer is the major product [Eq. (4.52)]. The *syn* selectivity can be as high as 10:1 (*syn*:*anti*) in the reaction of arabinose. It is argued that in this case, the allylindium intermediate coordinates with both the hydroxy and the carbonyl function as in **44**, leading to the *syn* adduct.

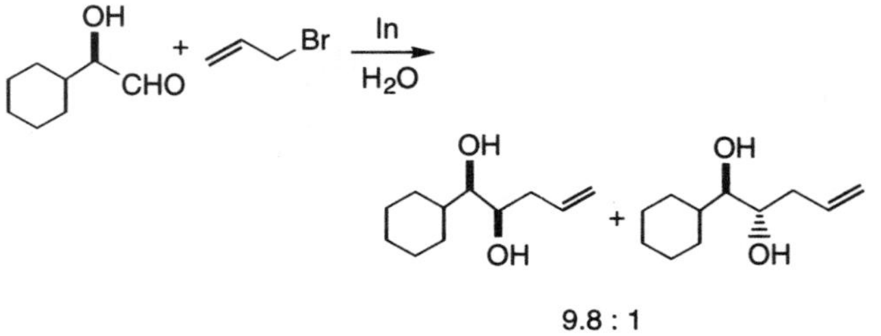

$$9.8:1 \tag{4.52}$$

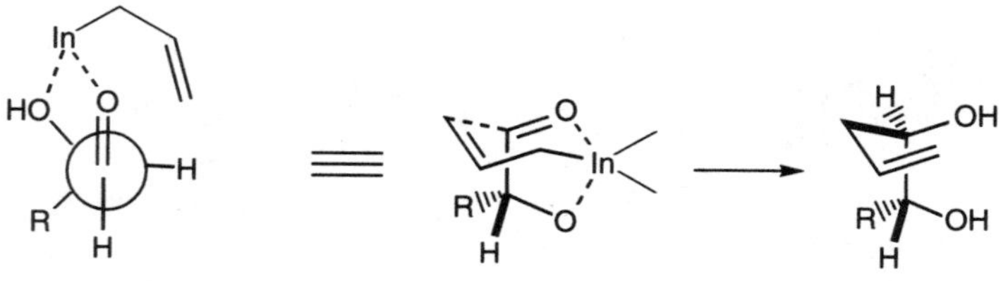

44   *syn-*

The same coordination is used to account for the observed *anti* preference in the allylation of β-hydroxybutanal (**45**) with allyl bromide/indium in water [Eq. (4.53)]. The intermediate **46** leads to the *anti* product **47**. In support of the intramolecular chelation model, it is found that if the hydroxy group is converted to the corresponding benzyl or *tert*-butyldimethylsilyl ether, the reaction is not stereoselective at all and gives nearly equal amounts of *syn* and *anti* products.

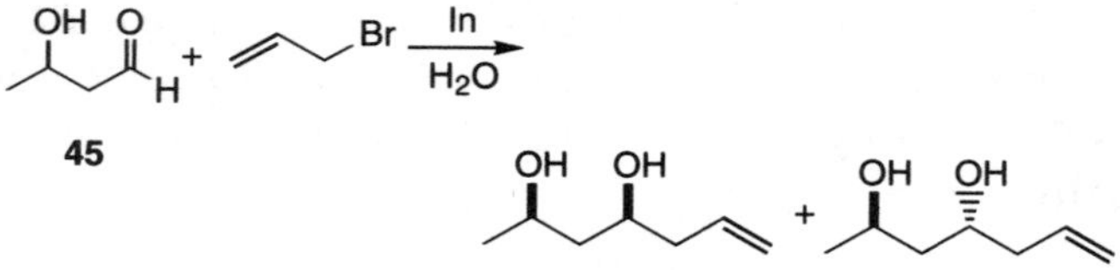

45

1 : 8.5

(4.53)

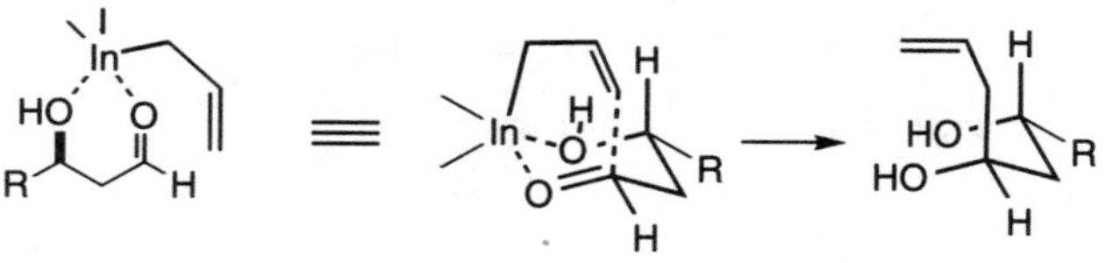

46   **47** *anti-*

It is possible to combine type A and type B situations in the coupling of a chiral aldehyde with a substituted allylic halide. Such is the case in the coupling of unprotected aldoses (e.g., glyceraldehyde) with cinnamyl bromide [Eq. (4.54)]. In such a coupling, two new stereogenic centers are created. It has been found that the *syn-,syn*-isomer **48** is formed preferentially. To account for the *syn-syn* stereochemistry, chelation of the allylindium species with the hydroxyaldehyde function with intramolecular attack through a cyclic transition state (**49**) is postulated. The stereochemistry of the adduct is then dependent on the geometry of the attacking allylindium species.

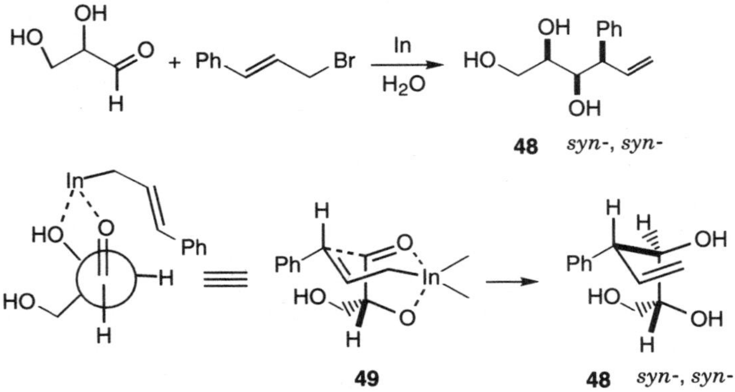

(4.54)

A variation of the type B situation is that the existing stereogenic center in the aldehyde can be farther away from the carbonyl group. Such an example can be found in Waldmann's studies of the diastereoselectivity of allylation reactions using the proline benzyl ester **50** as a chiral auxiliary to produce the α-hydroxyl amides **51** and **52**. The diastereoselectivity was ~4–5:1 (Scheme 4.8) (76). Separation of the diastereomers followed by reaction with methyl lithium produced the enantiomerically pure alcohol **53**.

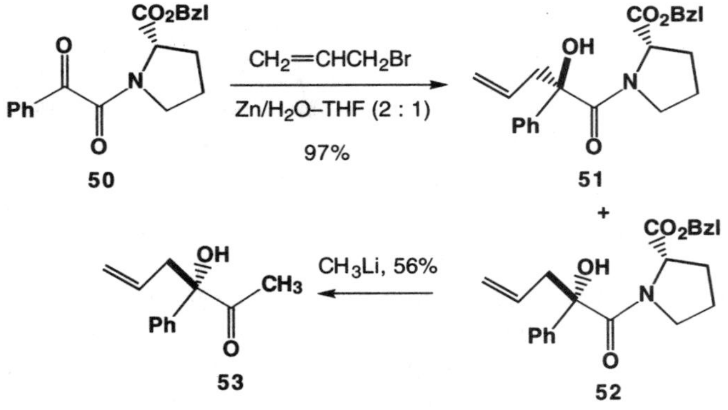

Scheme 4.8

4.1.9 Reaction Mechanism

Despite the successes of aqueous Barbier–Grignard reactions, the exact mechanism of these reactions is still largely unknown. Luche et al. (19) proposed that a free-radical pair process could be involved and has suggested that a radical derived from the halide attacks the carbonyl group. However, examination on a radical probe by Wilson et al. (46) did not give any definitive result (Scheme 4.9). Furthermore, the proposed free-radical mechanism also contradicts the chemoselectivity associated with the allylation of α,β-unsaturated carbonyl compounds, in which exclusive 1,2-addition products are obtained while free radicals tend to undergo conjugated additions.

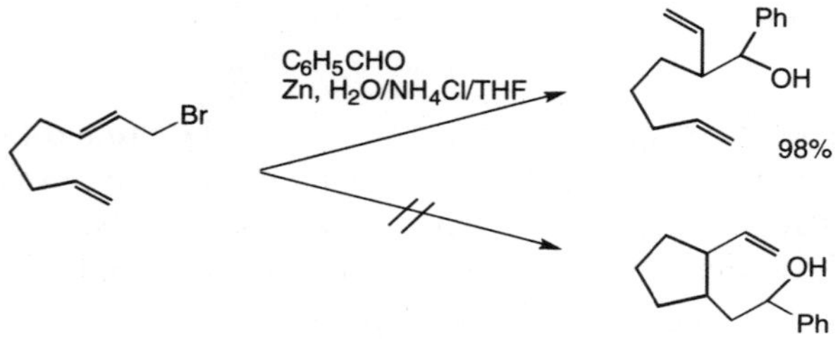

Scheme 4.9

Chan and Li have proposed a mechanism of a radical anion that coordinated on the metal surface (39a). In this mechanism, a single-electron-transfer (SET) process is involved (Scheme 4.10).

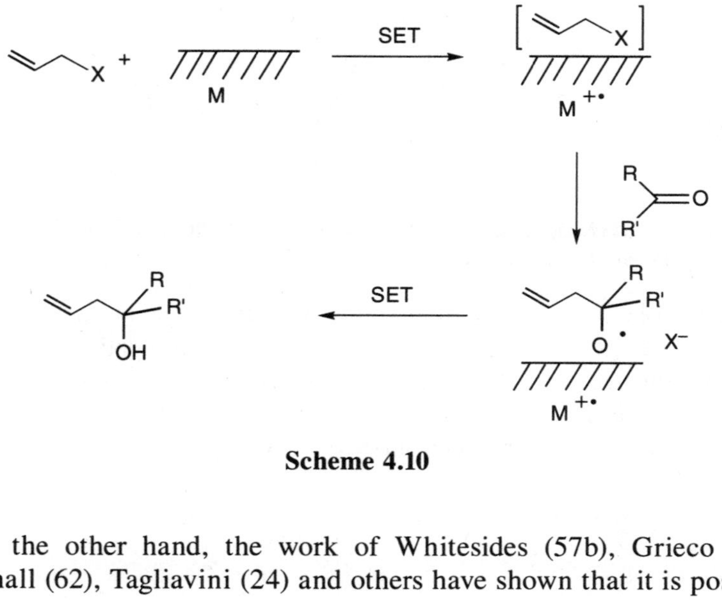

Scheme 4.10

On the other hand, the work of Whitesides (57b), Grieco (28), Marshall (62), Tagliavini (24) and others have shown that it is possible to carry out alkylation reactions in water with preformed allylmetal reagents (Scheme 4.11). Such results raise the possibility of a third mechanism involving an organometallic intermediate.

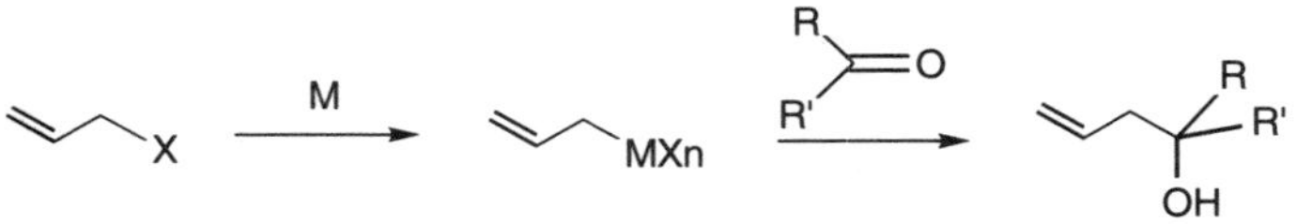

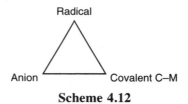

Scheme 4.11

It is possible that some element of truth could be found in each of the three proposed mechanisms. In any particular reaction, one mechanism might reflect more the reaction process than the others, depending on the reaction substrates and conditions. One might generalize the mechanism as a radical–anion–covalent (C–M) triad (Scheme 4.12). The actual mechanism may be located anywhere within the triad, while the exact location is determined by the reaction substrates, the metal used, and the reaction conditions. The three discrete species—of radical, anion, or covalent structure—may then represent the conceptually idealized situations.

Radical

Anion   Covalent C–M

Scheme 4.12

4.1.10 Metal-Mediated Olefination of Carbonyl Compounds (Wittig-Type Reactions)

Carbonyl olefination with methylmolybdenum reagents in aqueous media has been reported (77). In these cases, the molybdenum reagents were preformed in THF. The carbonyl compound, dissolved or suspended in water or ethanol/water mixture, or in ethanol, was added to the reagents to yield the corresponding alkenes:

$$PhCHO \;+\; MoOCl{=}CH_2 \xrightarrow[50\%]{} PhCH{=}CH_2 \qquad (4.55)$$

$$\text{PhCHO} \quad + \quad Cl_3Mo{=}CH_2 \xrightarrow[31\%]{} \text{PhCH}{=}CH_2 \qquad (4.56)$$

In both of these cases, considerable amount of the aldehyde was recovered, suggesting that the yield could be improved if the organometallic reagent was not presumably destroyed by water faster than the addition reaction. Further investigations of these reactions are clearly warranted.

4.2 METAL-MEDIATED CONJUGATED ADDITION

Although the allylation reactions in aqueous media were very successful, alkylation reactions in aqueous media with nonactivated alkyl halides usually failed to proceed, presumably due to the higher ionization potential of these compounds. However, Luche found that when zinc–copper couple was used, the less reactive alkyl halides reacted with conjugated carbonyl compounds and nitriles to give 1,4-addition products in good yields under sonication conditions [Eq. (4.57)] (78). A moderate diastereoselectivity was observed in these reactions where a mixture of diastereomers could be generated (79).

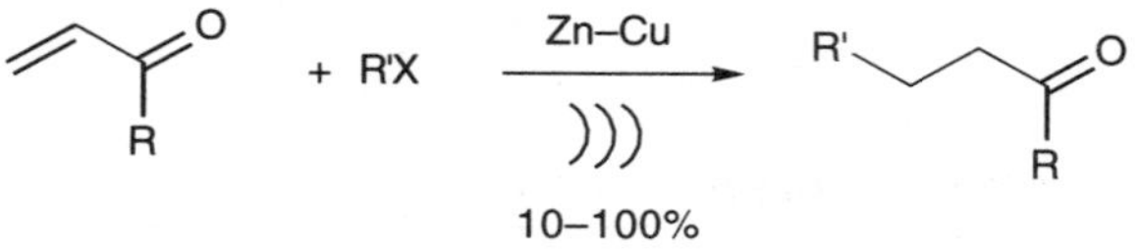

$$(4.57)$$

The reactivity of the halides followed the order of tertiary > secondary >> primary, and iodide > bromide (chlorides do not react). The preferred solvent system was aqueous ethanol. The process was suggested to proceed by a free-radical mechanism occurring on the metal surface under sonochemical conditions. Efforts to trap the intermediate intramolecularly gave only a very low yield of the cyclization product (80):

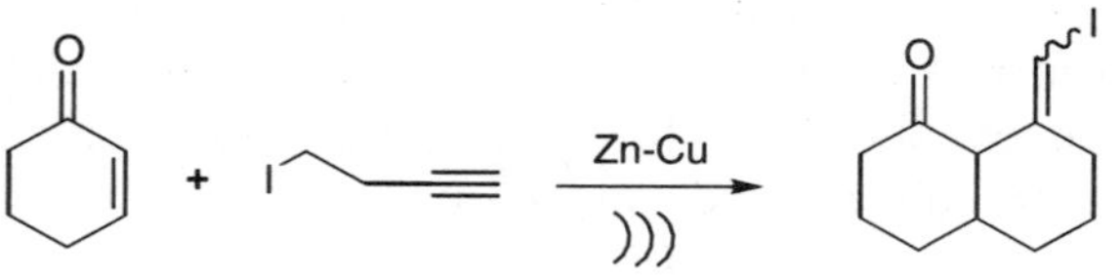

$$(4.58)$$

Similar additions also occurred on vinylphosphine oxides. When the optically active vinylphosphine oxide **54** was used, P-chiral alkylphosphine oxide **55** was obtained with retention of configuration (81):

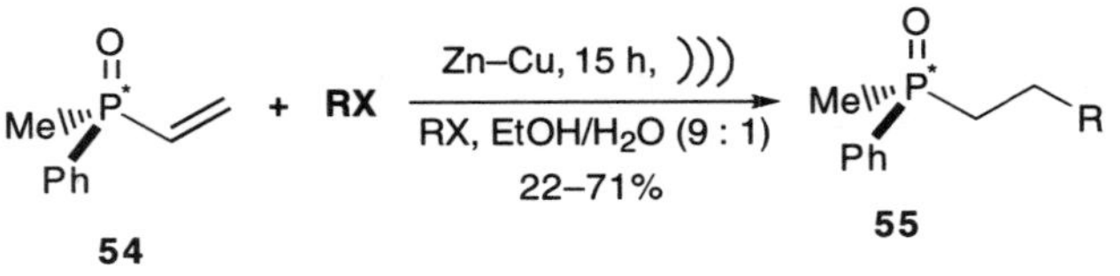

$$\text{(4.59)}$$

Recently, Giese studied the diastereoselectivity associated with such a conjugated addition in water [Eq. (4.60)] (Table 4.8) (82) and found *anti*-isomer to be the major product when the attacking radical was bulky.

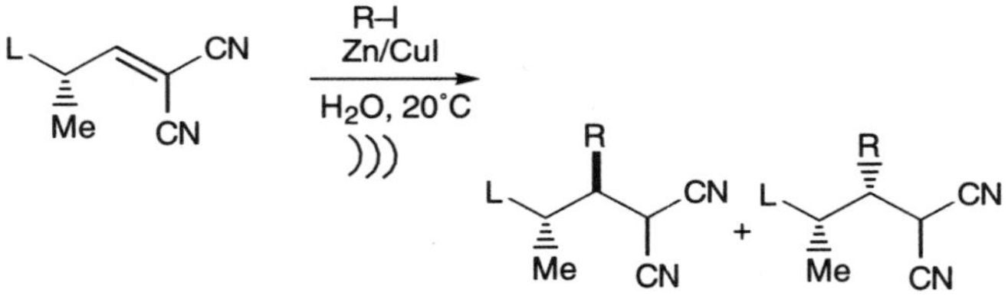

$$\text{(4.60)}$$

TABLE 4.8 Selectivity of the Addition

L	R	*anti/syn*	Yield (%)
t-Bu	*t*-Bu	>99:1	32
t-Bu	*i*-Pr	80:20	50
t-Bu	Et	50:50	20
t-Bu	Me	39:61	50
i-Pr	*t*-Bu	84:16	45
Et	*t*-Bu	72:28	61

The selectivity was explained using the argument that the more stable "A-strain" conformer of the alkene reacts more slowly with bulky alkyl radical than the less stable "Felkin–Anh" conformer, thus favoring the *anti* adducts in these cases (Scheme 4.13).

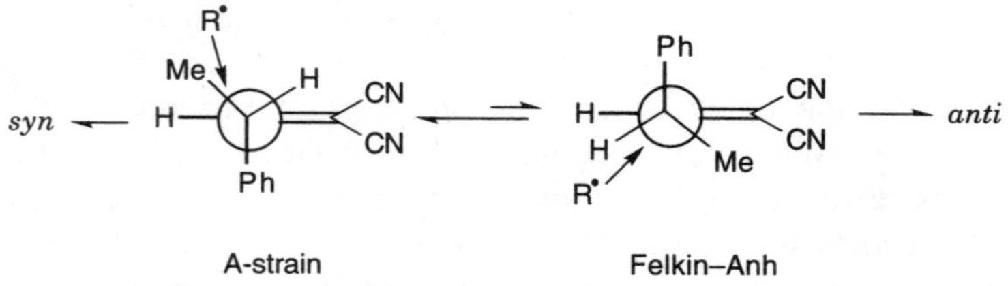

Scheme 4.13

4.3 WURTZ COUPLING

The homocoupling of allyl halides mediated by zinc in refluxing alcohol was one of the first metal-mediated reactions involving protic solvent (42). The yields of these reactions are generally modest. As shown by Li and Chan, the allyl halide **56** can undergo a Wurtz-type reaction mediated by zinc in water to give the homocoupled product (50):

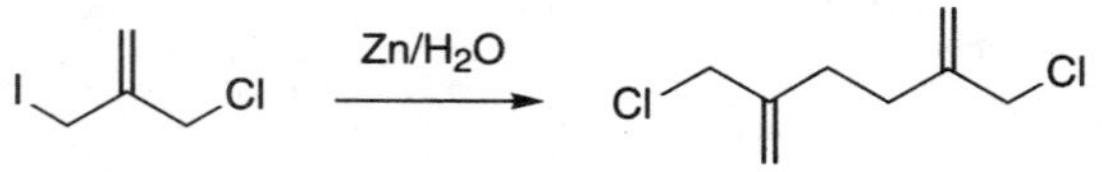

$$(4.61)$$

Recently, Tagliavini used the zinc-mediated Wurtz-type reaction of allyl bromide and haloorganotin compounds as a convenient route to allylstannanes and hexaaryldistannanes (83):

$$n\text{-RBr} + \text{Bu}_{4-n}\text{SnCl}_n \xrightarrow[27-90\ \%]{\text{THF/H}_2\text{O(NH}_4\text{Cl)/Zn}} \text{Bu}_{4-n}\text{SnR}_n$$

R = allyl- or allyl-like group
$n = 1,2$

$$(4.62)$$

4.4 PINACOL COUPLING

The coupling of carbonyl compounds (84) to give 1,2-diols, known as *pinacol coupling*, has been carried out in aqueous media. The use of a Zn–Cu couple to couple unsaturated aldehydes to pinacols was reported as early as 1892 (85). Early experiments also included the use of chromium and vanadium (86), as well as some ammoniacal TiCl$_3$-based reducing agents (87).

Recently, Clerici and Porta extensively studied the aqueous pinacol coupling reactions mediated by Ti(III). Aromatic ketones and aldehydes were homocoupled by TiCl₃ in aqueous solution under alkaline conditions (88). When this reagent was used under acidic conditions, aliphatic or aromatic ketones or aldehydes containing "activating" (strongly electron-withdrawing) groups reacted to give homocoupling products (89, 90). With nonactivated carbonyl compounds, cross-coupled products were obtained when the nonactivated carbonyl compounds were used in excess or as a solvent (91). The activating groups included CN, CHO, COMe, COOH, COOMe, and pyridyl. The mechanism of this reaction was suggested to be a radical process [Eq. (4.63)]. Very recently, Schwartz reported a stereoselective pinacol coupling with a cyclopentadienetitanium complex (92).

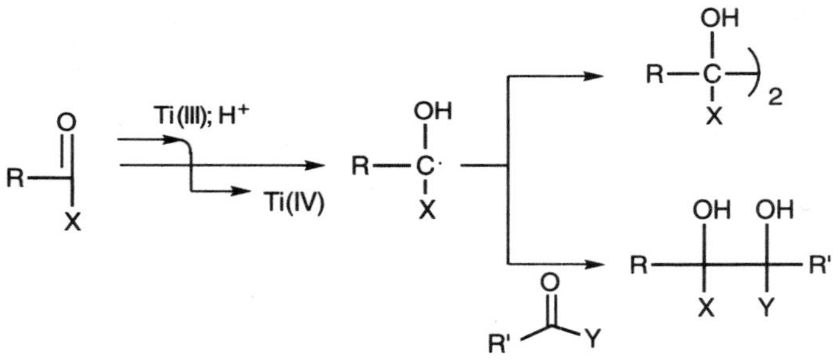

$$(4.63)$$

Cross-coupling reactions between α,β-unsaturated carbonyl compounds and acetone were carried out by using a Zn–Cu couple and ultrasonic radiation in an aqueous acetone suspension [Eq. (4.64)] (93). A similar radical mechanism was suggested. The large excess of acetone was intended to alleviate the self-coupling of the α,β-unsaturated substrates.

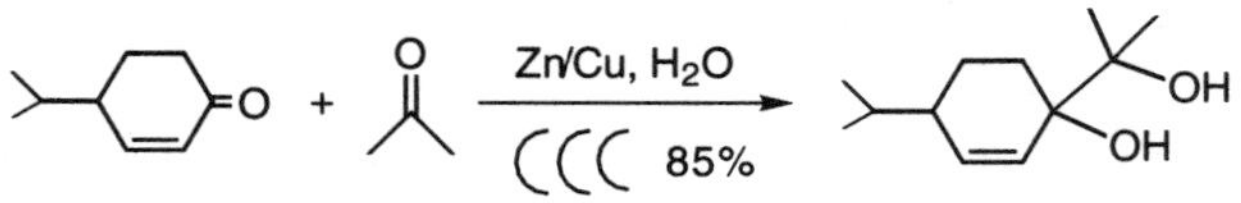

$$(4.64)$$

By using indium in aqueous ethanol, aldimines are reductively coupled to vicinal diamines [Eq. (4.65)]. The reaction is accelerated by the addition of ammonium chloride (94).

$$Ar_1 \overset{NAr_2}{\underset{H}{\bigg\Vert}} H \quad \xrightarrow[\substack{NH_4Cl \\ 0-100\%}]{In\text{--}H_2O\text{--}EtOH} \quad Ar_1 \overset{Ar_2HN \qquad NHAr_2}{\underset{H \qquad\quad H}{\bigg|\qquad\bigg|}} Ar_1 \qquad (4.65)$$

4.5 SYNTHETIC APPLICATIONS

By far, most of the synthetic applications of metal-mediated reactions have been Barbier-type reactions. Because of its superior reactivity, the indium-mediated reaction in water has found wider applications in natural product synthesis so far.

Synthetically, one of the most important features of carrying out organic reactions in water is that water-soluble hydroxyl-containing molecules can be used directly in reaction without the requirements of protection–deprotection processes. As such, the most important application of aqueous-medium Barbier–Grignard-type reactions in organic synthesis is in the field where extensive protection–deprotection processes are involved. A typical area is carbohydrate chemistry. Indeed, in 1991, Whitesides et al. reported the first application of aqueous-medium Barbier–Grignard reaction to carbohydrate synthesis through the use of tin in an aqueous–organic solvent mixture [Eq. (4.66)] (Table 4.9) (95).

$$
\begin{array}{c}
\text{CHO} \\
| \\
(\text{CHOH})_n \\
| \\
\text{CH}_2\text{OH}
\end{array}
\quad \xrightarrow[\text{H}_2\text{O/EtOH, ultrasound}]{\text{Sn, allyl bromide}} \quad
\begin{array}{c}
\text{CH}_2\text{OH} \\
| \\
(\text{CHOH})_{n-1} \\
\text{H}\!-\!\!-\!\text{OH} \\
\text{HO}\!-\!\!-\!\text{H} \\
| \\
\text{CH}_2 \\
| \\
\text{CH}\!=\!\text{CH}_2
\end{array}
\;+\;
\begin{array}{c}
\text{CH}_2\text{OH} \\
| \\
(\text{CHOH})_{n-1} \\
\text{HO}\!-\!\!-\!\text{H} \\
\text{H}\!-\!\!-\!\text{OH} \\
| \\
\text{CH}_2 \\
| \\
\text{CH}\!=\!\text{CH}_2
\end{array}
\qquad (4.66)
$$

The adducts were converted to higher carbon aldoses by ozonolysis of the deprotected polyols followed by suitable derivatization (Scheme 4.14). The reaction showed a higher diastereoselectivity when there was a hydroxyl group present at C_2. However, no reaction was ob-

TABLE 4.9 Tin-Mediated Allylation of Aldoses

Carbohydrate	Diastereoselectivity	Yield (%)
D-Erythrose	4.0:1	52
D-Ribose	3.5:1	65
D-Arabinose	5.5:1	85
D-Glucose	6.5:1	70
D-Mannose	6.0:1	90
2-Deoxy-D-glucose	1.5:1	26
2-NAc-D-glucose	—	0
2-NAc-D-mannose	—	0

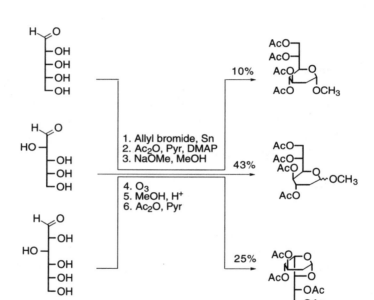

Scheme 4.14

served under the reaction conditions when there was an *N*-acetyl group present at the C$_2$ position.

Shortly afterward, Chan and Li reported (96) an efficient synthesis of (+)-3-deoxy-D-*glycero*-D-*galacto*-nonulosonic acid (KDN, **57**) (Scheme 4.15), using the indium-mediated alkylation reaction in water. A similar synthesis of 3-deoxy-D-*manno*-octulonate (KDO, **58**) led primarily to the undesired diastereomer. However, through the disruption of the newly generated stereogenic center (97), they completed a formal

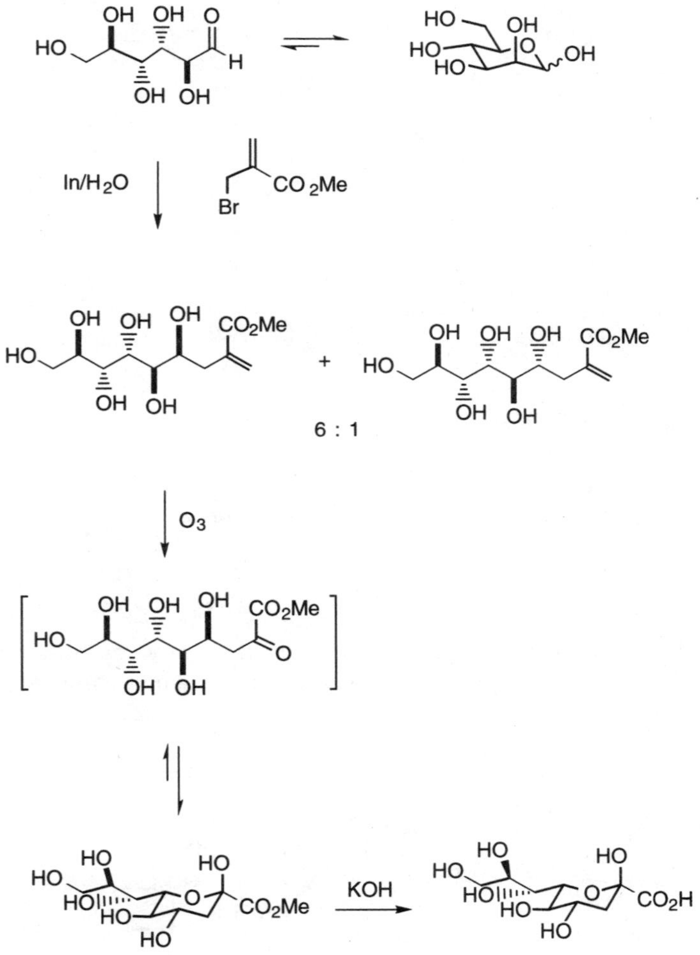

Scheme 4.15

synthesis (98) of KDO (Scheme 4.16). In contrast to the tin-mediated reactions, the indium-mediated reaction also occurred on a substrate with an *N*-acetyl group present at C_2, which later led to the efficient synthesis of *N*-acetyl-neuraminic acid (99).

Later, Whitesides et al. reported the synthesis of *N*-acetylneu-raminic acid (**59**) (57b, 100), as well as other sialic acid derivatives (**60** and **61**) based on a similar strategy. The use of indium is essential for the carbon–carbon bond-formation step in these sialic acid syntheses.

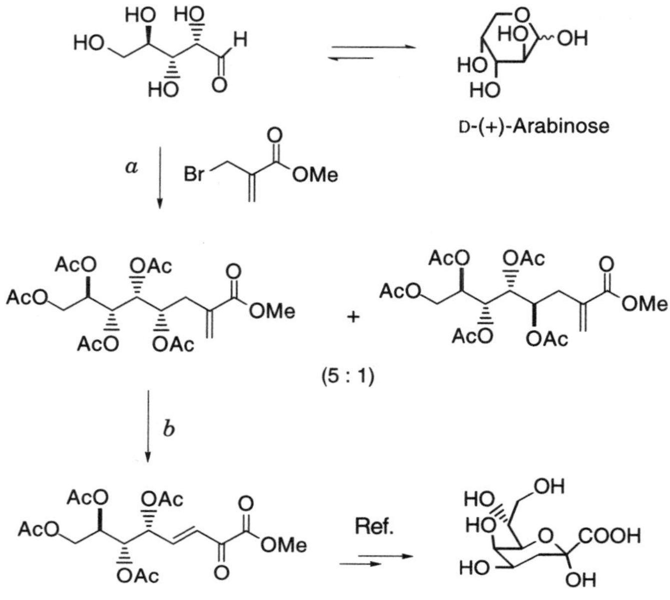

D-(+)-Arabinose

(5 : 1)

Ref.

58 (+)-KDO

Scheme 4.16 (*a*) In/H$_2$O, vigorous stirring, then Ac$_2$O/pyridine/DMAP, 79%; (*b*) O$_3$/methylene chloride $-78°$C to RT, then column chromatography, 67%.

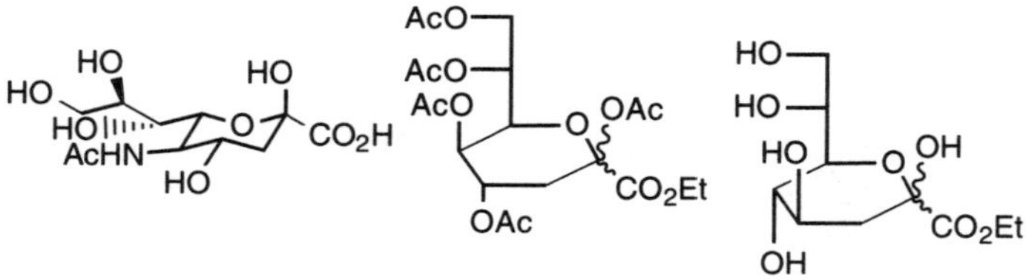

59   **60**   **61**

KDO was synthesized via indium mediated allylation of 2,3:4,5-di-*O*-isopropylidene-D-arabinose (Scheme 4.17) (101). In this case, the desired product became the major product due to the protection of the α-hydroxyl group.

Recently, Chan et al. further shortened the already concise sialic acid synthesis to two steps through the indium-mediated reaction of α-(bromomethyl)acrylic acid with sugars (Scheme 4.18). Both KDN and *N*-acetylneuraminic acid have been synthesized in this way (58).

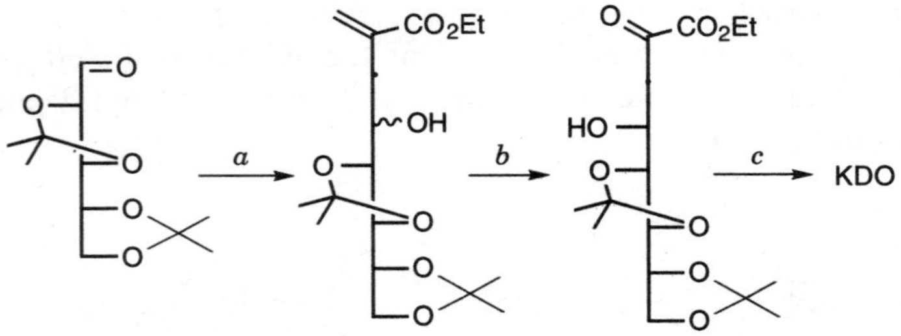

erythro : *threo* = 2 : 1

Scheme 4.17 (*a*) In/ethyl α-(bromomethyl)acrylate, 10% formic acid, aqueous MeCN, 61%; (*b*) O₃, MeOH, −78°C to RT, 92%; (*c*) TFA, NH₄OH, 55%.

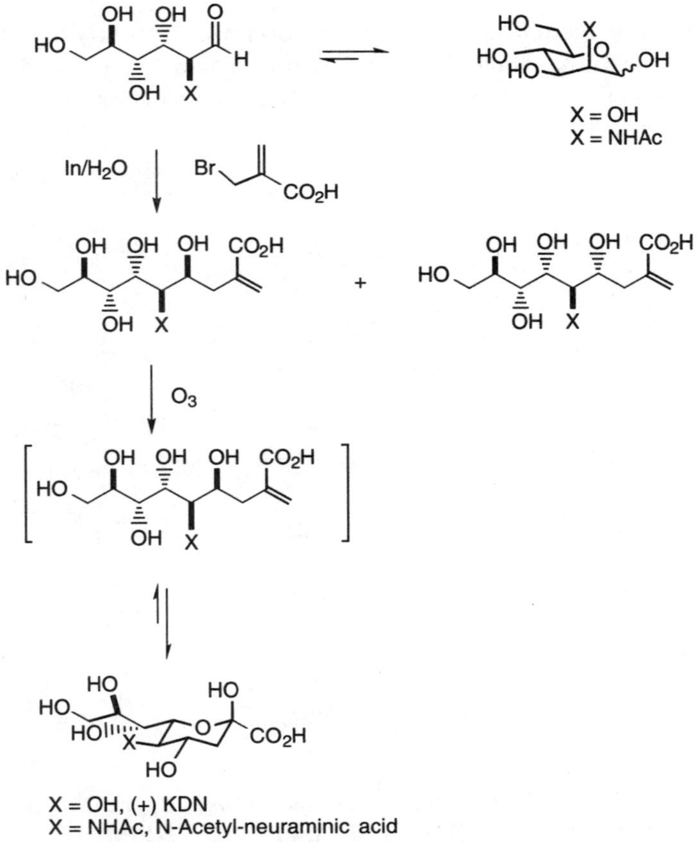

X = OH
X = NHAc

X = OH, (+) KDN
X = NHAc, N-Acetyl-neuraminic acid

Scheme 4.18

Phosphonic acid analog **62** and **63** of both KDN and *N*-acetylneu-raminic acid have been synthesized using the indium-mediated coupling of the lower carbohydrates with dimethyl 3-bromopropenyl-2-phospho-nate (**64**) in water (70).

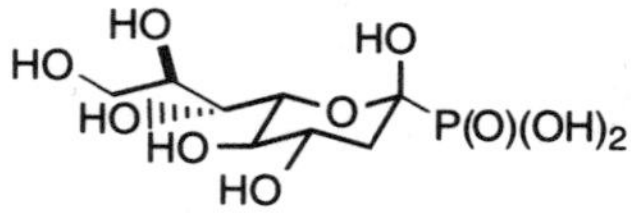

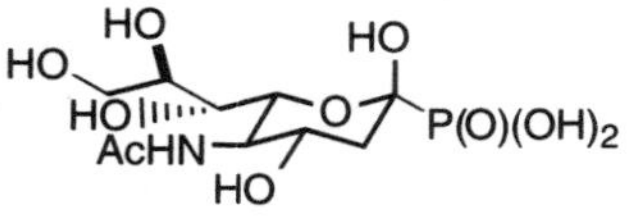

The indium-mediated allylation reaction was recently applied by Schmid et al. to the elongation of the carbon chain of carbohydrates in forming higher analogs (Scheme 4.19) (102) and to deoxy sugars (**65–68**) (103).

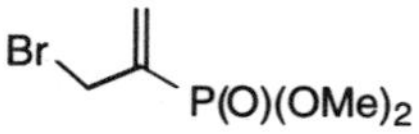

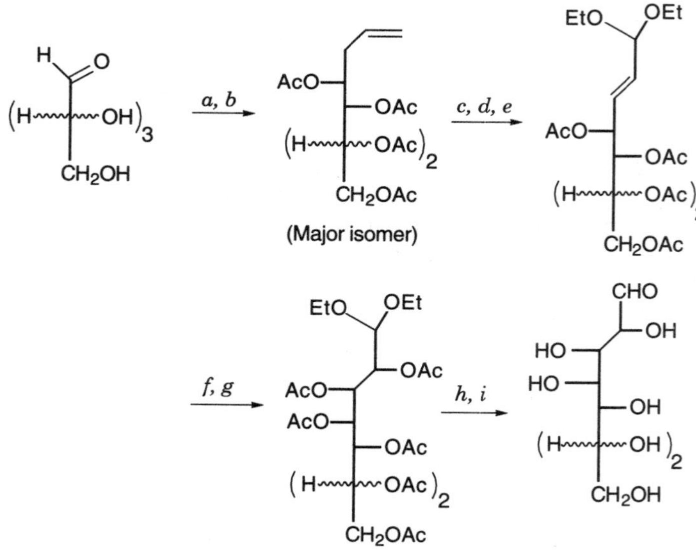

Scheme 4.19   (*a*) In/allyl bromide, ultrasound; (*b*) Ac$_2$O/pyridine/DMAP; (*c*) OsO$_4$, KIO$_4$; (*d*) TBAF; (*e*) H$^+$/HC(OEt)$_3$; (*f*) OsO$_4$, NMMNO; (*g*) Ac$_2$O/pyridine/DMAP; (*h*) NaOMe/MeOH; (*i*) H$^+$.

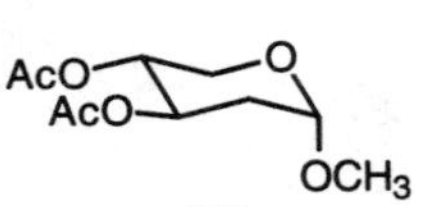

65 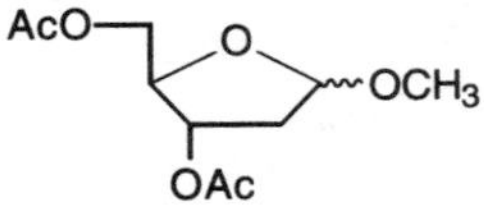   66

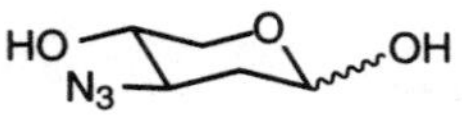

67 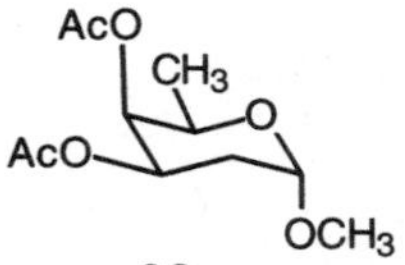   68

Loh reported the reaction of the glucose-derived aldehyde with allyl bromide mediated by indium [Eq. (4.67)] (104). The reaction again

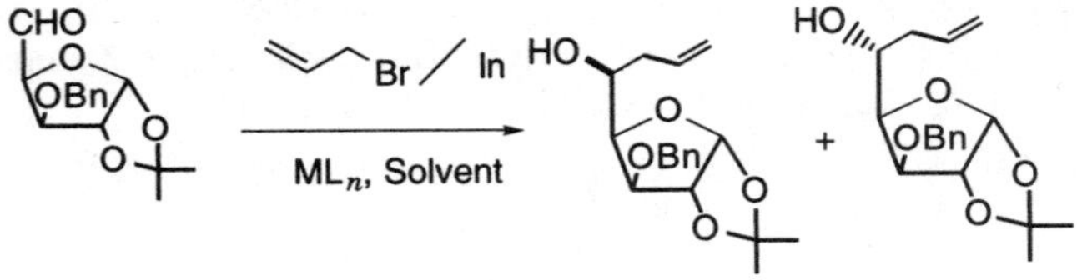

(4.67)

H_2O (10 h)	41:59
H_2O–THF(4:1) (10 h)	24:76
H_2O–DMF(4:6) (2 h)	17:83
H_2O–DMF(4:6)-Yb(OTf)$_3$ (1 h)	6:94

gives a nonchelation product as the major diastereomer. The use of an organic cosolvent increases the diastereoselectivity. The addition of ytterbium trifluoromethanesulfonate [Yb(OTf)$_3$] enhances both the reactivity and the diastereoselectivity.

The aqueous-medium Barbier–Grignard-type reaction has also been used in other syntheses besides carbohydrates. Chan and Li used the zinc-mediated allylation as a key step in a total synthesis of (+)-muscarine **(69)** (Scheme 4.20) (105). The strategy was based on the observation that the diastereoselectivity of the allylation reaction in water can be reversed through protection of the α-hydroxyl group.

As reported by Li et al., enolizable 1,3-dicarbonyl compounds undergo efficient carbonyl allylation reactions in aqueous media [Eq. (4.68)] (106). The reaction is general, and a variety of 1,3-dicarbonyl compounds have been alkylated successfully using allyl

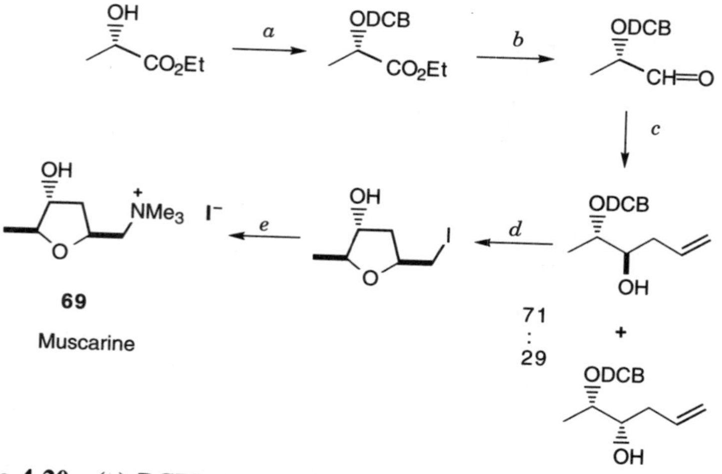

Scheme 4.20 (a) $DCBBr/Ag_2O/Et_2O/reflus/6$ h (90%); (b) $DIBAL-H/E_2O/$ $-78°C/2$ h; (c) $CH_2=CHCH_2Br/Zn/H_2O/NH_4Cl/3$ h (85%), 2 steps); (d) $I_2/$ $CH_3CN/0°C/3$ h (85%); (e) $NMe_3/EtOH/80°C/4$ h (60%).

bromide or allyl chloride in conjunction with either tin or indium. Zinc can also be used, although it is not as effective (107). This reaction can be used readily for the synthesis cyclopentane derivatives [Eq. (4.69)] (Table 4.10) (108).

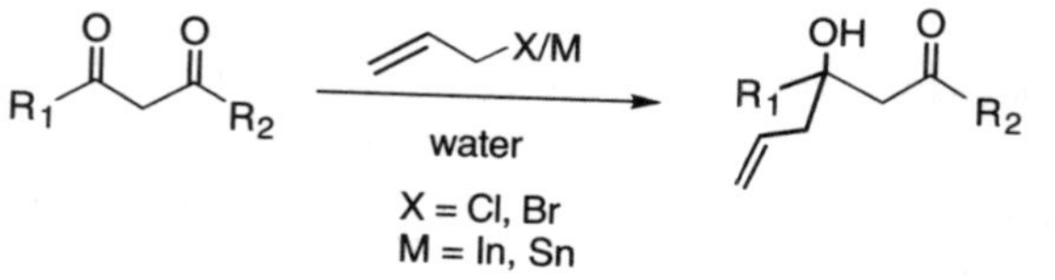

$$(4.68)$$

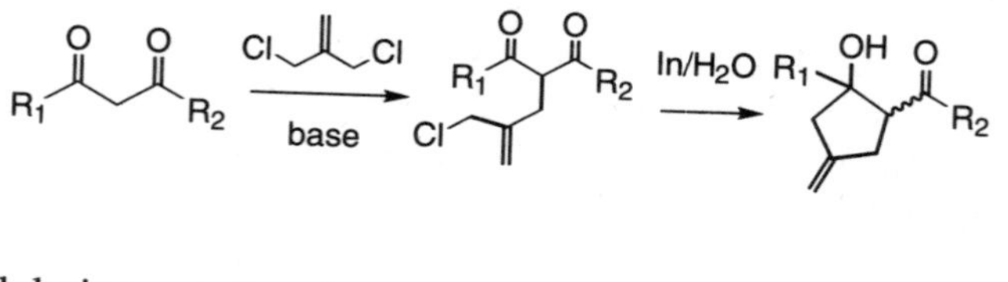

$$(4.69)$$

The allylation reaction in water also could be used to prepare α,α-difluorohomoallylic alcohols, reported by Burton et al., from *gem*-difluoro allyl halides (109).

TABLE 4.10 [3 + 2] Annulation in Water to Cyclopentanoids

Substrate	Intermediate (Yield)	Product [Yield (%)]
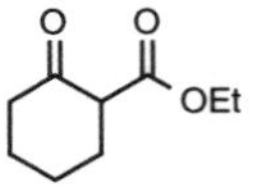	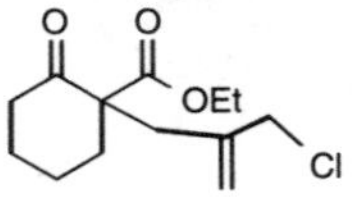 (65)	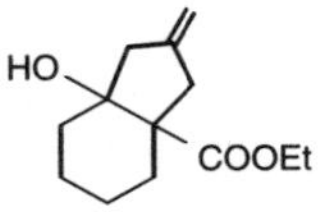 (98)
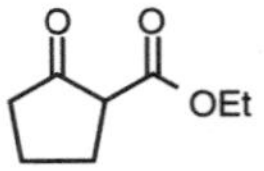	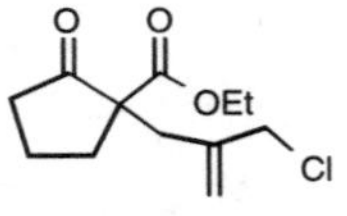 (63)	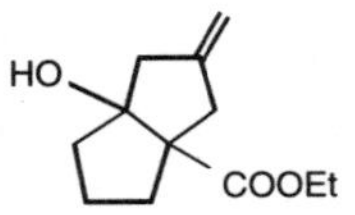 (98)
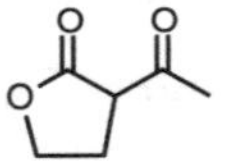	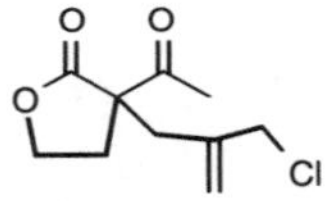 (60)	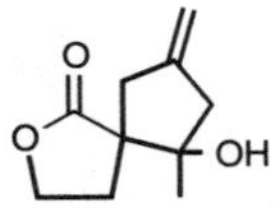 (quant.)
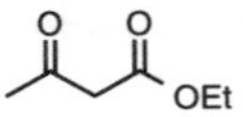	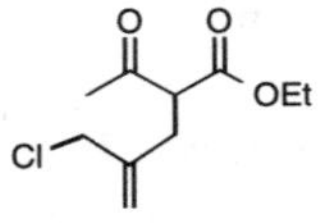 (60)	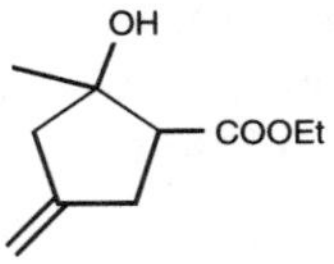 (92)
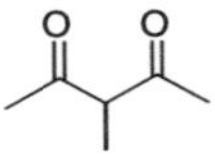	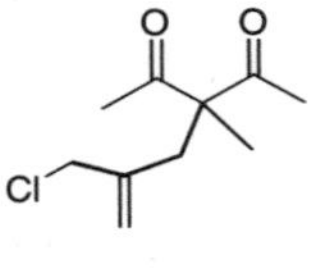 (60)	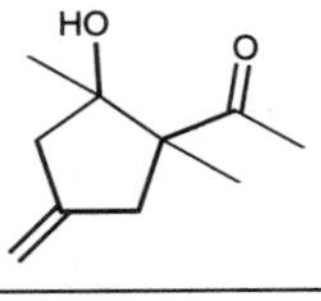 (90)

An intriguing application of aqueous Barbier-type reaction in organic syntheses is a novel carbocycle ring enlargement methodology developed by Li et al. (110). Traditionally, for two-atom ring expansions, the photochemical method of [2 + 2] cyclization–decyclization is the most successful (111). The [2 + 2] cycloaddition of an acetylenic ester to an enamine of a cyclic ketone and subsequent opening of the annulated cyclobutene moiety is another useful method for two-carbon ring expansion (112). Other $n + 2$ ring expansions include the 1,3-migration of allylic alcohols or ethers (113), or the aldol-type condensations (114).

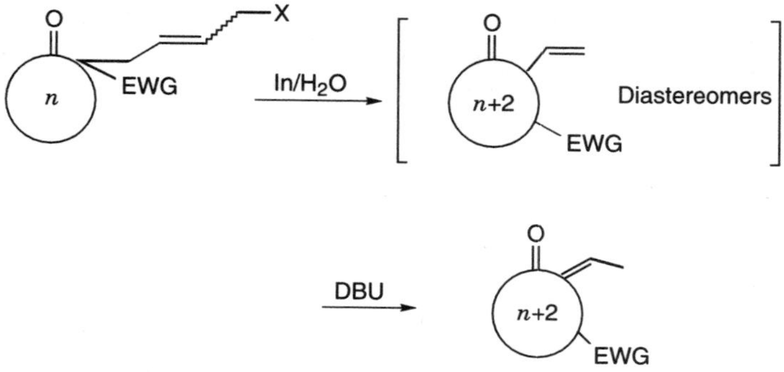

Scheme 4.21

By using the indium-mediated Barbier-type reaction in water, 5-, 6-, 7-, 8-, and 12-membered rings are enlarged by two carbon atoms into 7-, 8-, 9-, 10-, and 14-membered ring derivatives, respectively (Scheme 4.21) (Table 4.11). The use of water as a solvent was found to be critical for the success of the reaction. Similar ring expansion in organic solvents was not successful.

A 5–6 fused-ring system was also efficiently transformed into a 5–8 fused-ring compound (Scheme 4.22). Such a ring expansion method

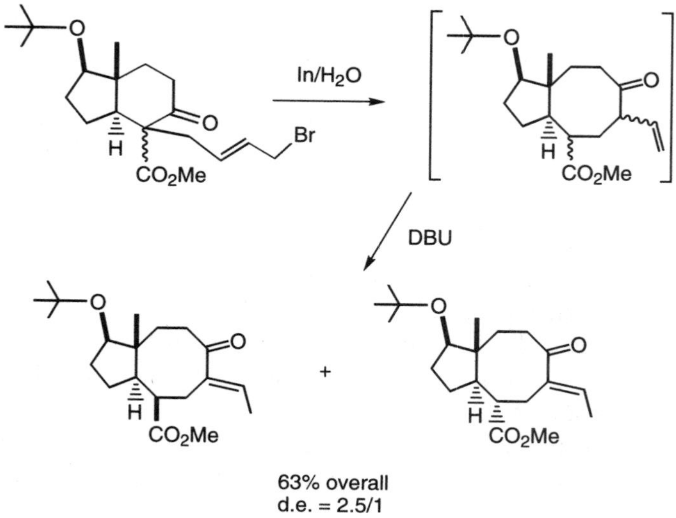

63% overall
d.e. = 2.5/1

Scheme 4.22

TABLE 4.11 Carbocycle Enlargement to Medium and Large Rings

Substrate	Intermediate	Product [Overall Yield (%)]
		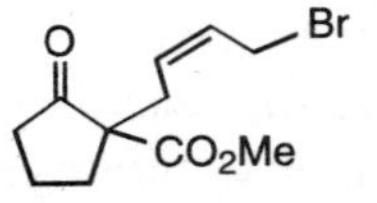 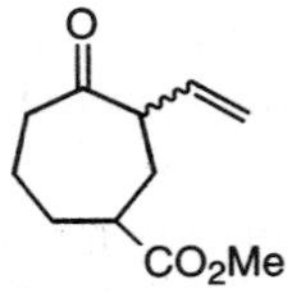 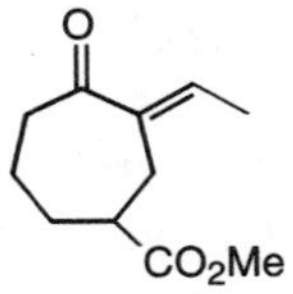 (50)
		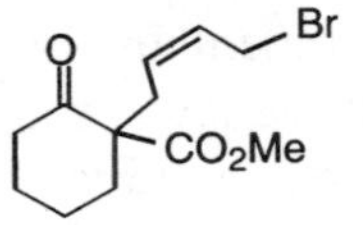 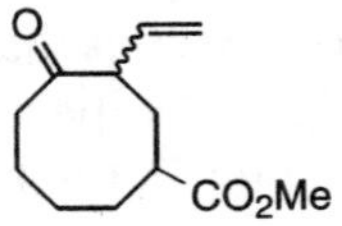 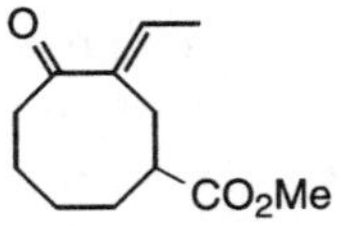 (51)
		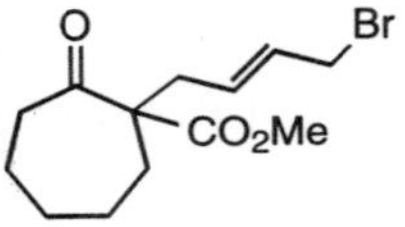 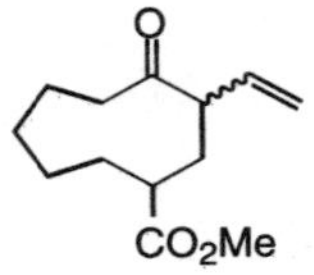 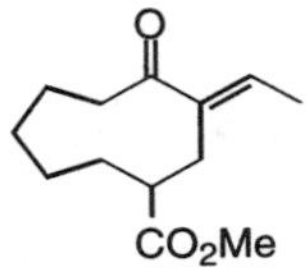 (71)
		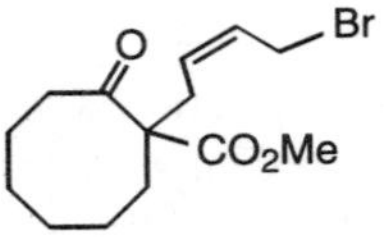 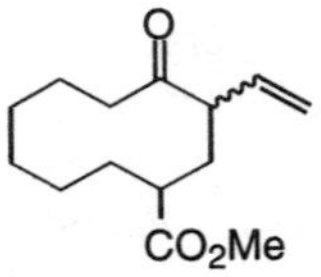 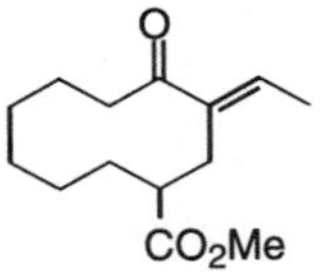 (50)
		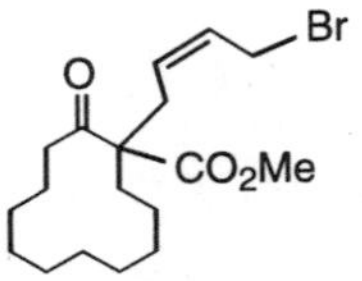 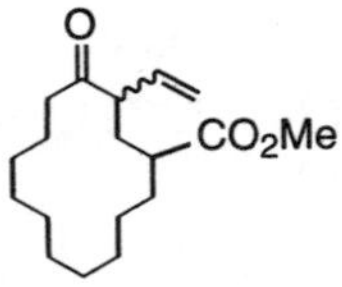 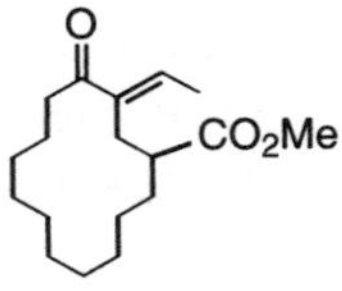 (49)
		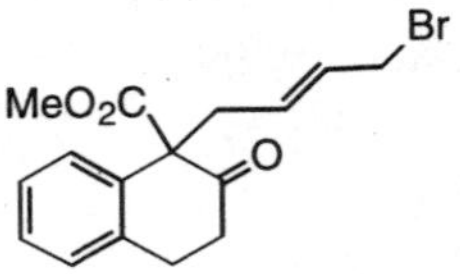 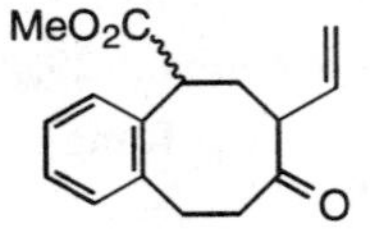 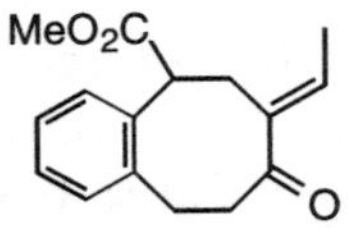 (51)

could find extensive applications, especially in the syntheses of medium and large ring carbocycles.

4.6 CONCLUSION

In spite of their usefulness in synthesis as demonstrated in this chapter, Barbier–Grignard-type reactions in water are still in the infancy stage.

Since their history is only a decade old, the full synthetic potential of such reactions remains to be explored. The reactions need to be extended to nonactivated alkyl halides. More active and more efficient mediators than indium and other currently used metals may have to be discovered. The factors that affect the stereochemistry of the reactions need to be better understood in order to improve the stereoselectivity. Enantioselective synthesis through such aqueous-medium reactions has yet to be explored and certainly will be a major development in this area in the future. Thus, it can be expected that the use of aqueous solvent for carrying out Barbier–Grignard-type and other metal-mediated reactions will be an important chapter in the history of organometallic reactions.

REFERENCES

1. Frankland, E. *Ann.* **71**, 171 (1849).

2. For a recent monograph, see Blomberg, C. "The Barbier Reaction and Related One-Step Processes," in *Reactivity and Structure: Concepts in Organic Chemistry*, Hafner, K.; Lehn, J. M.; Rees, C. W.; von Ragué Schleyer, P.; Trost, B. M.; Zahradník, R., eds., Springer-Verlag, 1993.

3. Reformatsky, A. *Ber.* **20**, 1210 (1887).

4. Barbier P. *Compt. Rend.* **128**, 110 (1899).

5. Grignard, V. *Compt. Rend.* **130**, 1322 (1990).

6. Peters, W. *Ber.* **38**, 2567 (1905).

7. Sisido, K.; Takeda,Y.; Kinugawa, Z. *J. Am. Chem. Soc.* **83**, 538 (1961); Sisido, K.; Kozima, S.; Hanada, T. *J. Organomet. Chem.* **9**, 99 (1967); Sisido, K.; Kozima, S. *J. Organomet. Chem.* **11**, 503 (1968).

8. Nosek, J. *Collect. Czech. Chem. Commun.* **29**, 597 (1964).

9. Killinger, T. A.; Boughton, N. A.; Runge, T. A.; Wolinsky, J. *J. Organomet. Chem.* **124**, 131 (1977).

10. For general reviews on organic reactions in aqueous medium, see (a) Li, C. J. *Tetrahedron* **52**, 5643 (1996); (b) Chan, T. H.; Isaac, M. B. *Pure & Appl. Chem.* **68**, 919 (1996); (c) Li, C. J. *Chem. Rev.* **93**, 2023 (1993).

11. Nokami, J.; Otera, J.; Sudo, T.; Okawara, R. *Organometallics* **2**, 191 (1983).

12. Nokami, J.; Wakabayashi, S.; Okawara, R. *Chem. Lett.* 869 (1984).

13. Zhou, J. Y.; Chen, Z. G.; Wu, S. H. *J. Chem. Soc., Chem Commun.* 2783 (1994).

14. Uneyama, K.; Kamaki, N.; Moriya, A.; Torii, S. *J. Org. Chem.* **50**, 5396 (1985).

15. Wu, S. H.; Huang, B. Z.; Zhu, T. M.; Yiao, D. Z.; Chu, Y. L. *Act. Chim. Sinica.* **48**, 372 (1990); ibid. **45**, 1135 (1987).

16. Uneyama, K.; Matsuda, H.; Torii, S. *Tetrahedron Lett.* **25**, 6017 (1984).

17. Mandai, T.; Nokami, J.; Yano, T.; Yoshinaga, Y.; Otera, J. *J. Org. Chem.* **49**, 172 (1984).

18. Petrier, C.; Einhorn, J.; Luche, J. L. *Tetrahedron Lett.* **26**, 1449 (1985).

19. Einhorn, C.; Luche, J. L. *J. Organomet. Chem.* **322**, 177 (1987).

20. Masuyama, Y.; Takahara, T. P.; Kurusu, Y. *Tetrahedron Lett.* **30**, 3437 (1989); Masuyama, Y.; Nimura, Y.; Kurusu, Y. *Tetrahedron Lett.* **32**, 225 (1991).

21. Sati, M.; Sinou, D. *Tetrahedron Lett.* **32**, 2025 (1991).

22. Masuyama, Y.; Kishida, M.; Kurusu, Y. *J. Chem. Soc. Chem. Commun.* 1405 (1995).

23. Wu, S. H.; Huang, B. Z.; Gao, X. *Synth. Commun.* **20**, 1279 (1990).

24. Boaretto, A.; Marton, D.; Tagliavini, G.; Gambaro, A. *J. Organomet. Chem.* **286**, 9 (1985).

25. Boaretto, A.; Marton, D.; Tagliavini, G. *J. Organomet. Chem.* **297**, 149 (1985).

26. Furlani, D.; Marton, D.; Tagliavini, G.; Zordan, M.; *J. Organomet. Chem.* **341**, 345 (1988).

27. Hachiya, I.; Kobayashi, S. *J. Org. Chem.* **58**, 6958 (1993).

28. Grieco, P. A.; Bahsas, A. *J. Org. Chem.* **52**, 1378 (1987).

29. Nokami, J.; Tamaoka, T.; Ogawa, H.; Wakabayash, S. *Chem. Lett.* 541 (1986).

30. Uneyama, K.; Ueda, K.; Torii, S. *Chem. Lett.* 1201 (1986).

31. Talaga, P.; Schaeffer, M.; Benezra, C.; Stampf, J. L. *Synthesis* 530 (1990).

32. Zhou, J. Y.; Lu, G. D.; Wu, S. H. *Synth. Commun.* **22**, 481 (1992).

33. House, H. O., *Modern Synthetic Reactions*, 2nd ed., Benjamin, New York, 1972.

34. Mukaiyama,T. *Isr. J. Chem.* **24**, 162 (1984).

35. Mukaiyama,T. *Org. React.* **28**, 203 (1982).

36. Stevens, R. W.; Iwasawa, N.; Mukaiyama, T. *Chem. Lett.* 1459 (1982).

37. Lubineau, A. *J. Org. Chem.* **51**, 2143 (1986); Lubineau, A.; Meyer, E. *Tetrahedron* **44**, 6065 (1988).

38. Kobayashi, S.; Hachiya, I. *Tetrahedron Lett.* **33**, 1625 (1992); Hachiya, I.; Kobayashi, S. *J. Org. Chem.* **58**, 6958 (1993).

39. (a) Chan, T. H.; Li, C. J.; Wei, Z. Y. *J. Chem. Soc., Chem. Commun.* 505 (1990); (b) Zhou, J. Y.; Jia, Y.; Shao, Q. Y.; Wu, S. H. *Synth. Commun.* **26**, 769 (1996).

40. Li, C. J.; Lee, M. C.; Chan, T. H. (in preparation).

41 Killinger, T. A.; Boughton, N. A.; Runge, T. A.; Wolinsky, J. *J. Organomet. Chem.* **124**, 131 (1977).

42. Nosek, J.; *Collect. Czech. Chem. Commun.* **29**, 597 (1964).

43. Petrier, C.; Luche, J. L. *J. Org. Chem.* **50**, 910 (1985).

44. Einhorn, C.; Luche, J. L. *J. Organomet. Chem.* **322**, 177 (1987).

45. Mattes, H.; Benezra, C. *Tetrahedron Lett.* **26**, 5697 (1985).

46. Wilson, S. R.; Guazzaroni, M. E. *J. Org. Chem.* **54**, 3087 (1989).

47. Kunz, T.; Reissig, H. U. *Liebigs Ann. Chem.* 891 (1989).

48. Chan, T. H.; Li, C. J.; Lee, M. C.; Wei, Z. Y. *Can. J. Chem.* **72**, 1181 (1994).

49. Chan, T. H.; Li, C. J. *Organometallics* **9**, 2649 (1990).

50. Li, C. J.; Chan, T. H. *Organometallics* **10**, 2548 (1991).

51. Trost, B. M.; King, S. A. *J. Am. Chem. Soc.* **112**, 408 (1990).

52. Oda, Y.; Matsuo, S.; Saito, K. *Tetrahedron Lett.* **33**, 97 (1992).

53. Oda, Y. private communication. See also ref. 52.

54. Cripps, H. N.; Kiefer, E. F. *Org. Syn.* **42**, 12–14 (1962).

55. Durant, A.; Delplancke, J. L.; Winand, R.; Reisse, J. *Tetrahedron Lett.* **36**, 4257 (1995).

56. Li, C. J.; Chan, T. H. *Tetrahedron Lett.* **32**, 7017 (1991).

57. (a) Araki, S.; Jin, S. J.; Idou, Y.; Butsugan, Y. *Bull. Chem. Soc. Jpn.* **65**, 1736 (1992); (b) Kim, E.; Gordon, D. M.; Schmid, W.; Whitesides, G. M. *J. Org. Chem.* **58**, 5500 (1993).

58. Chan, T. H.; Lee, M. C. *J. Org. Chem.* **60**, 4228 (1995).

59. Li, C. J. *Tetrahedron Lett.* **36**, 517 (1995).

60. Chen, D. L.; Li, C. J. *Tetrahedron Lett.* **37**, 295 (1996).

61. Isaac, M. B.; Chan, T. H. *Tetrahedron Lett.* **36**, 8957 (1995).

62. Marshall, J. A.; Hinkle, K. W. *J. Org. Chem.* **60**, 1920 (1995).

63. Wada, M.; Ohki, H.; Akiba, K. Y. *Bull. Chem. Soc. Jpn.* **63**, 1738 (1990); *J. Chem. Soc., Chem. Commun.* 708 (1987).

64. Katritzky, A. R.; Shobana, N.; Harris, P. A. *Organometallics* **11**, 1381 (1992); Katrizky, A. R.; Allin, S. M. *Synth. Commun.* **25**, 2751 (1995).

65. Minato, M.; Tsuji, J. *Chem. Lett.* 2049 (1988).

66. Li, C. J.; Chan, T. H. unpublished results.

67. Kauffmann, T.; Abeln, R.; Wingbermuhle, D. *Angew. Chem., Int. Ed. Engl.* **23**, 730 (1984).

68. Yanagisawa, A.; Inoue, H.; Morodome, M.; Yamamoto, H. *J. Am. Chem. Soc.* **115**, 10356 (1993).

69. Isaac, M. B., Ph.D. thesis, McGill University, 1996.

70. Chan, T. H.; Xin, Y .C. *J. Chem. Soc., Chem. Commun.* 905 (1996).

71. Isaac, M. B.; Chan, T. H. *Tetrahedron Lett.* **36**, 8957 (1995).

72. Uneyama, K.; Nanbu, H.; Torii, S. *Tetrahedron Lett.* **27**, 2395 (1986).

73. Chan, T. H.; Li, C. J. *Can. J. Chem.* **70**, 2726 (1992).

74. Cram, D. J.; Kopecky, K. R. *J. Am. Chem. Soc.* **81**, 2748 (1959); Reetz, M. T. *Angew. Chem., Int. Ed. Eng.* **23**, 556 (1984); Midland, M. M.; Koops, R. W. *J. Org. Chem.* **55**, 5058 (1990).

75. Paquette, L.; Mitzel, T. M. *J. Am. Chem. Soc.* **118**, 1931 (1996).

76. Waldmann, H. *Synlett.* 627 (1990).

77. Kauffmann, T.; Fiegenbaum, P.; Wieschollek, R. *Angew. Chem., Int. Ed. Engl.* **23**, 531 (1984).

78. (a) Petrier, C.; Dupuy, C.; Luche, J. L. *Tetrahedron Lett.* **27**, 3149 (1986); Luche, J. L.; (b) Allavena, C. *Tetrahedron Lett.* **29**, 5369 (1988); (c) Dupuy, C.; Petrier, C.; Sarandeses, L. A.; Luche, J. L. *Synth. Commun.* **21**, 643 (1991).

79. Giese, B.; Damm, W.; Roth, M.; Zehnder, M.; *Synlett.* 441 (1992); Erdmann, P.; Schafer, J.; Springer, R.; Zeitz, H. G.; Giese, B. *Helv. Chim. Act.* **75**, 638 (1992).

80. Luche, J. L.; Allavena, C.; Petrier, C.; Dupuy, C. *Tetrahedron Lett.* **29**, 5373 (1988).

81. Pietrusiewicz, K. M.; Zablocka, M. *Tetrahedron Lett.* **29**, 937 (1988).

82. Roth, M.; Damm, W.; Giese, B. *Tetrahedron Lett.* **37**, 351 (1996).

83. von Gyldenfeldt, F. Marton, D.; Tagliavini, G. *Organometallics* **13**, 906 (1994).

84. For recent reviews, see Kahn, B. E.; Rieke, R. D. *Chem. Rev.* **88**, 733 (1988); Pons, J. M.; Santelli, M. *Tetrahedron* **44**, 4295 (1988).

85. Grinder, *Ann. Chim. Phys.* **26**, 369 (1892).

86. Conant, J. B.; Cutter, H. B. *J. Am. Chem. Soc.* **48**, 1016 (1926).

87. Karrer, P.; Yen, Y.; Reichstein, I. *Helv. Chim. Acta* **13**, 1308 (1930).

88. Clerici, A.; Porta, O. *Tetrahedron Lett.* **23**, 3517 (1982).

89. Clerici, A.; Porta, O. *J. Org. Chem.* **47**, 2852 (1982).

90. Clerici, A.; Porta, O.; Riva, M. *Tetrahedron Lett.* **22**, 1043 (1981); Reference is cited therein.

91. Clerici, A.; Porta, O. *J. Org. Chem.* **48**, 1690 (1983); Clerici, A.; Porta, O. *Tetrahedron* **39**, 1239 (1983); Clerici, A.; Porta, O.; Zago, P. *Tetrahedron* **42**, 561 (1986); Clerici, A.; Porta, O. *J. Org. Chem.* **54**, 3872 (1989).

92. Barden, M. C.; Schwartz, J. *J. Am. Chem. Soc.* **118**, 5484 (1996).

93. Delair, P.; Luche, J. L. *J. Chem. Soc., Chem. Commun.* 398 (1989).

94. Kalyanam, N; Rao V. G. *Tetrahedron Lett.* **34**, 1647 (1993).

95. Schmid, W.; Whitesides, G. M. *J. Am. Chem. Soc.* **113**, 6674 (1991).

96. Chan, T. H.; Li, C. J. *J. Chem. Soc., Chem. Commun.* 747 (1992).

97. Chan, T. H.; Li, C.J. 203rd National Meeting of the American Chemical Society, SanFrancisco, CA; April 1992, Abstract ORGN435.

98. Dondoni, A.; Merino, P.; Orduna, J. *Tetrahedron Lett.* **32**, 3247 (1991).

99. Li, C. J. Ph.D. thesis, McGill University, 1992.

100. Gordon, D. M.; Whitesides, G. M. *J. Org. Chem.* **58**, 7937 (1993).

101. Gao, J.; Harter, R.; Gordon, D. M.; Whitesides, G. M. *J. Org. Chem.* **59**, 3714 (1994).

102. Prenner, R. H.; Binder, W. H.; Schmid, W. *Liebigs Ann. Chem.* 73 (1994).

103. Binder, W. H.; Prenner, R. H.; Schmid, W. *Tetrahedron* **50**, 749 (1994).

104. Wang, R.; Lim, C. M.; Tan, C. H.; Lim, B. K.; Sim, K. Y.; Loh, T. P. *Tetrahedron: Asymmetry* **6**, 1825 (1995).

105. Archibald, S. C.; Hoffmann, R. W. *Chemtracts-Organic Chemstry* **6**, 194 (1993).

106. Li, C. J.; Lu, Y. Q. *Tetrahedron Lett.* **36**, 2721 (1995).

107. Ahonen, M.; Sjöholm, R. *Chem. Lett.* 341 (1995).

108. Li, C. J.; Lu, Y. Q. *Tetrahedron Lett.* **37**, 471 (1996).

109. Yang, Z. Y.; Burton, D. J. *J. Org. Chem.* **56**, 1037 (1991).

110. Li, C. J.; Chen, D. L.; Lu, Y. Q.; Haberman, J. X.; Mague, J. T. *J. Am. Chem. Soc.* **118**, 4216 (1996); Li, C. J. 211th National ACS Meeting, New Orleans, March 24–28, 1996.

111. For reviews, see de Mayo, P. *Pure Appl. Chem.* **9**, 597 (1964); de Mayo, P *Acc. Chem. Res.* **4**, 41 (1971); for reviews on synthetic applications, see Crimmins, M. T. *Chem. Rev.* **88**, 1453 (1988); Schuster, D. I.; Lem, G.; Kaprinidis, N. A. *Chem. Rev.* **93**, 3 (1993); De Keukeleire, D.; He, S. L. *Chem. Rev.* **93**, 359 (1993); Schreiber, S. L. *Science* **227**, 857 (1985); Winkler, J. D.; Bowen, C. M.; Liotta, F. *Chem. Rev.* **95**, 2003 (1995).

112. For examples, see Lin, M. S.; Snieckus, V. *J. Org. Chem.* **36**, 645 (1971); Reinhoudt, D. N.; Verboom, W.; Visser, G. W.; Trompenaars, W. P.; Harkema, S.; van Hummel, G. J. *J. Am. Chem. Soc.* **106**, 1341 (1984).

113. Evans, D. A.; Golob, A. M. "[3,3]-Sigmatropic Rearrangements of 1,5-Diene Alkoxide. The Powerful Accelerating Effects of the Alkoxide Substituents," *J. Am. Chem. Soc.* **97**, 4765 (1975).

114. Nakashita, Y.; Hesse, M. "The Carbon Zip Reaction," *Angew. Chem., Int. Ed. Engl.* **20**, 1021 (1981).

CHAPTER 5

TRANSITION-METAL-CATALYZED REACTIONS

Nature always makes the best of possible things.
—Aristotle (384–322 B.C.)*

As in the case of organometallic reactions, transition-metal-catalyzed reactions were perceived for a long time to be only workable ones in the absence of water. While many transition-metal-catalyzed reactions are indeed sensitive toward water, recent studies have shown that numerous other transition-metal-catalyzed reactions can be carried out equally or more effectively by using water as solvent (1).

The use of transition-metal reagents in aqueous solvent systems offers the same advantages for a wide variety of chemical reactions ranging from large-scale industrial processes to laboratory organic synthesis. For large-scale chemical manufacturing, the advantage lies in simplifying product isolation and recycling the catalyst. In 1973, the first attempt was reported on a hydrogenation reaction of olefins in aqueous solution with a water-soluble phosphine ligand (2). No exceptional property was observed with the catalyst. However, the experiment showed that water did not interfere with the soft catalytic system. According to the hard/soft–acid/base theory, water is considered

*T. P. Kiernan, ed., *Aristotle Dictionary*, Philosophical Library, New York, 1962.

a "hard" base, whereas the intermediates in many transition-metal-catalyzed reactions are often considered to have a "soft" character (1).

5.1 WATER-SOLUBLE LIGANDS

There are two property-determining parts in nearly all transition-metal catalysts: the metal itself and the ligands. While the nature of a catalytic reaction is determined by the central metal, the ligand often affects the selectivity, for example, the enantioselectivity. In some cases, the choice of ligand also influences the outcome of a reaction. For catalysis in an aqueous medium, generally a water-soluble catalyst having water-soluble ligands is involved. Such a system could allow the easy separation of product from catalyst by simple phase separation, and the catalyst solution can be recycled easily. At the beginning of this chapter, we will first summarize water-soluble ligands that have been prepared and used in the literature. Since most catalysis involves the use of phosphine ligands, water-soluble ligands have been compiled into three tables: achiral water-soluble phosphine ligands (Table 5.1), chiral water-soluble phosphine ligands (Table 5.2), and other water-soluble ligands (Table 5.3).

TABLE 5.1 Achiral Water-Soluble Phosphine Ligands

Ligand (Abbreviation)	Ref.
$Ph_2P(m\text{-}C_6H_4SO_3Na)$	3
$P(m\text{-}C_6H_4SO_3Na)_3$	4
$Ph_2P(p\text{-}C_6H_4SO_3Na)$	5
$Ph_2P(m\text{-}C_6H_4CO_2Na)$	6
$Ph_2P(p\text{-}C_6H_4CO_2Na)$	6
$Ph_2P(o\text{-}C_6H_4CO_2Na)$	7
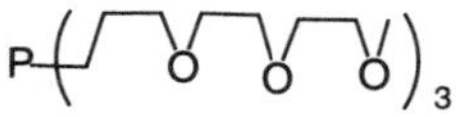	8
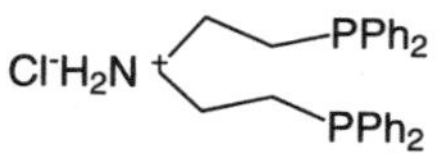	9
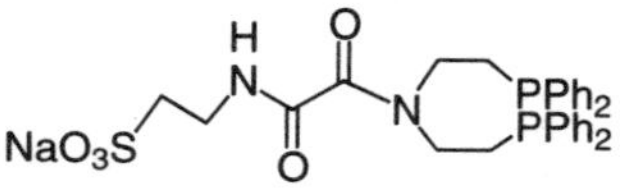	9

TABLE 5.1 Achiral Water-Soluble Phosphine Ligands (*continued*)

Ligand (Abbreviation)	Ref.
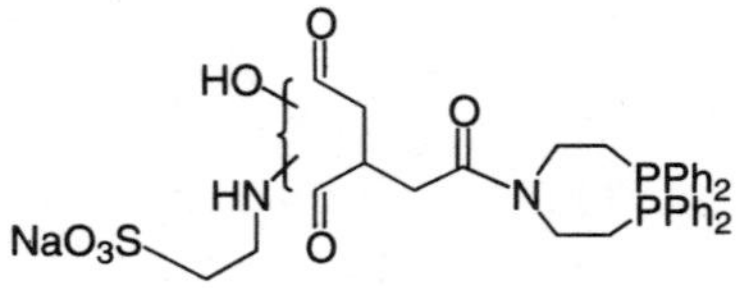	9
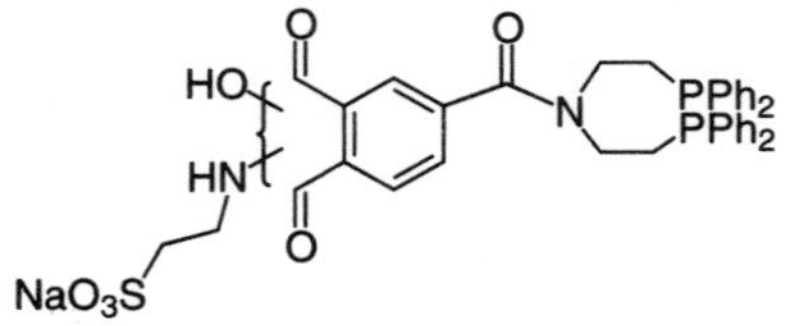	9
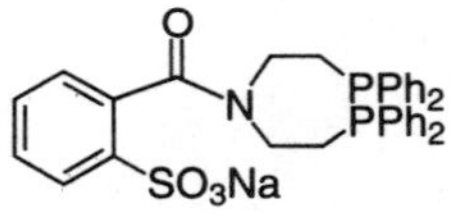	9
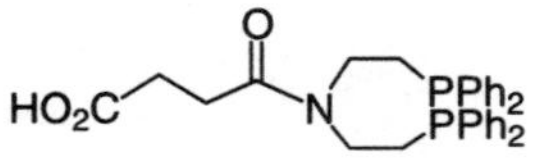	9
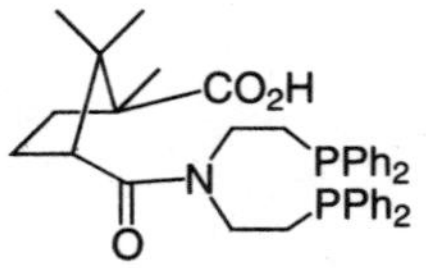	9
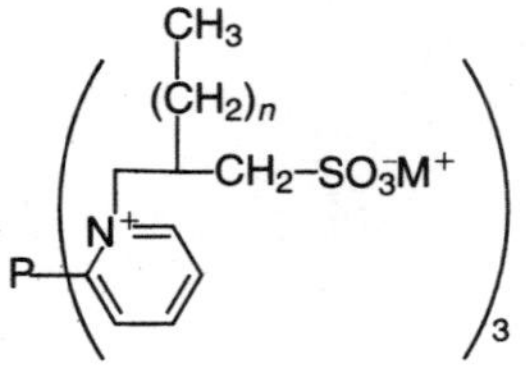	10
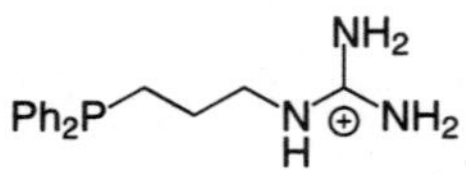	11
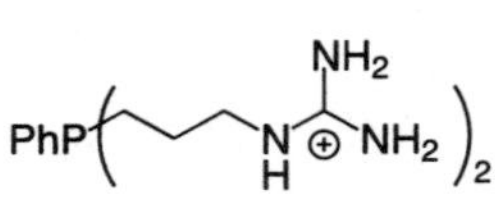	11
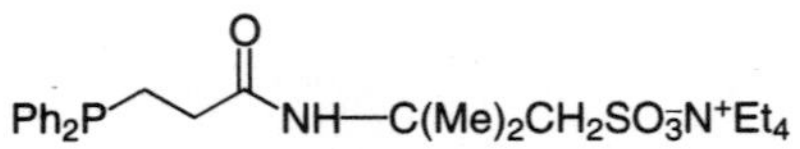	12

TABLE 5.1 Achiral Water-Soluble Phosphine Ligands (*continued*)

Ligand (Abbreviation)		Ref.
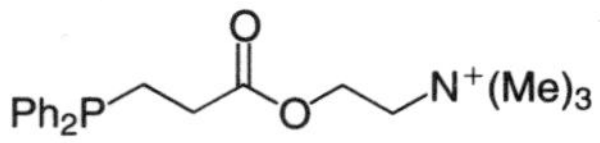 Ph_2P—CH$_2$CH$_2$C(=O)—O—CH$_2$CH$_2$—N$^+$(Me)$_3$		12
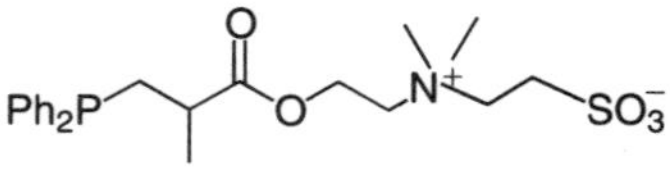 Ph_2P—CH(CH$_3$)—C(=O)—O—CH$_2$CH$_2$—N$^+$(Me)$_2$—CH$_2$CH$_2$—SO$_3^-$		12
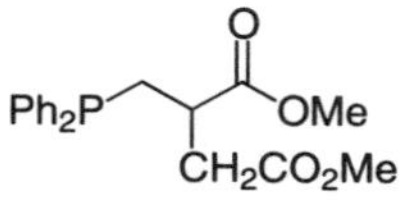 Ph_2P—CH(CH$_2$CO$_2$Me)—C(=O)—OMe		12
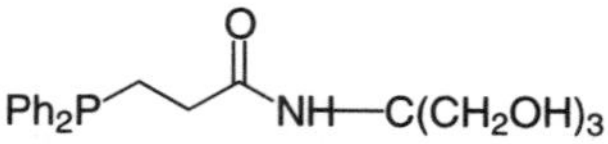 Ph_2P—CH$_2$CH$_2$C(=O)—NH—C(CH$_2$OH)$_3$		12
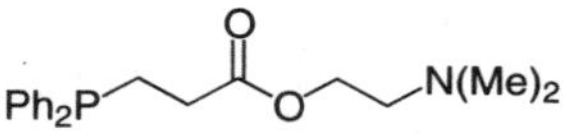 Ph_2P—CH$_2$CH$_2$C(=O)—O—CH$_2$CH$_2$—N(Me)$_2$		12
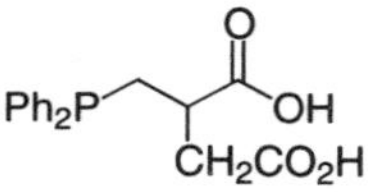 Ph_2P—CH(CH$_2$CO$_2$H)—C(=O)—OH		12
P(PhCH$_2$CH$_2$PhSO$_3$Na)$_3$		13
PhP(m-PhSO$_3$Na)$_2$	(tppds)	14
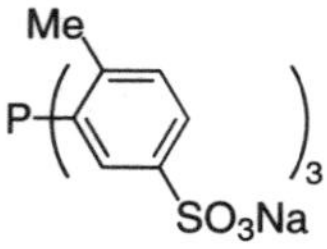 Me / SO$_3$Na substituted P(aryl)$_3$		14
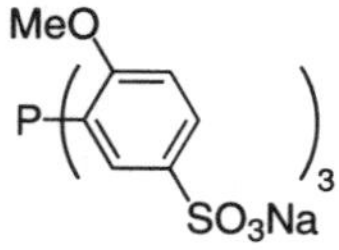 MeO / SO$_3$Na substituted P(aryl)$_3$		14
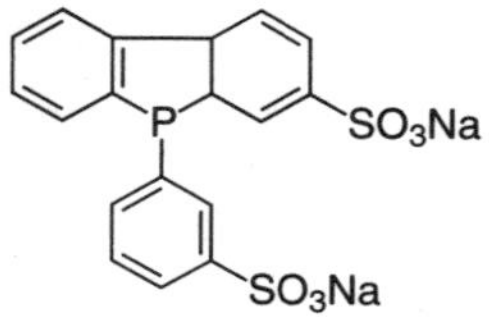 dibenzophosphole with SO$_3$Na and m-C$_6$H$_4$SO$_3$Na substituents		14

TABLE 5.1 Achiral Water-Soluble Phosphine Ligands (*continued*)

Ligand (Abbreviation)		Ref.
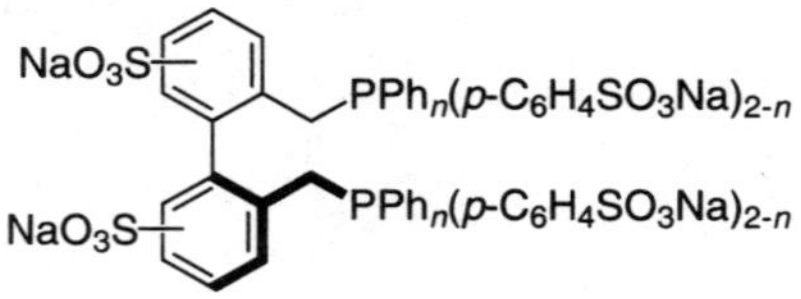 NaO$_3$S—[]—PPh$_n$(p-C$_6$H$_4$SO$_3$Na)$_{2-n}$ / NaO$_3$S—[]—PPh$_n$(p-C$_6$H$_4$SO$_3$Na)$_{2-n}$	(bisbis)	15
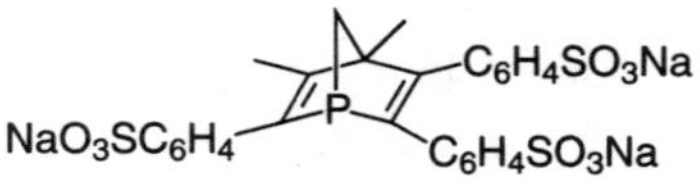 NaO$_3$SC$_6$H$_4$— ... —C$_6$H$_4$SO$_3$Na / C$_6$H$_4$SO$_3$Na	(norbos)	16
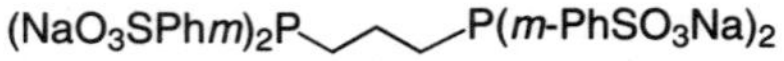 (NaO$_3$SPhm)$_2$P⌒⌒P(m-PhSO$_3$Na)$_2$		17
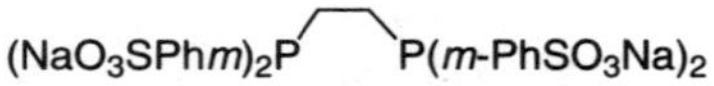 (NaO$_3$SPhm)$_2$P⌒P(m-PhSO$_3$Na)$_2$		18
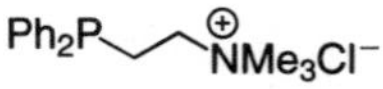 Ph$_2$P⌒N⁺Me$_3$Cl⁻	(amphos)	19
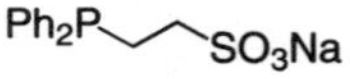 Ph$_2$P⌒SO$_3$Na		20
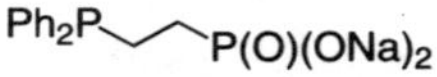 Ph$_2$P⌒P(O)(ONa)$_2$		20
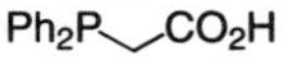 Ph$_2$P⌒CO$_2$H		21
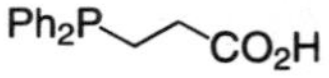 Ph$_2$P⌒CO$_2$H		22
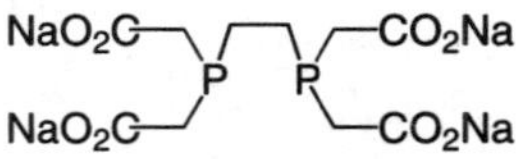 NaO$_2$C⌒P⌒P⌒CO$_2$Na / NaO$_2$C⌒ ⌒CO$_2$Na		23
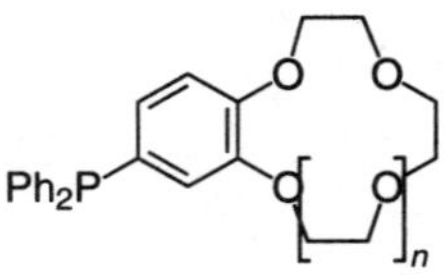 Ph$_2$P—[crown ether] $_n$	$n = 1,2,3,4$	24
P(CH$_2$CH$_2$OH)$_3$		25
P(CH$_2$OH)$_3$		26
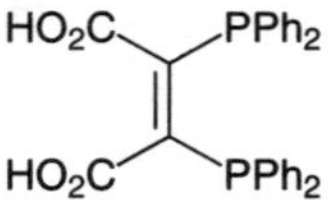 HO$_2$C⌒PPh$_2$ / HO$_2$C⌒PPh$_2$		27
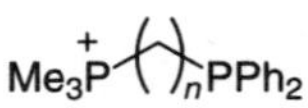 Me$_3$P⁺—()$_n$—PPh$_2$	(phosphos)	28

TABLE 5.1 Achiral Water-Soluble Phosphine Ligands (*continued*)

Ligand (Abbreviation)		Ref.
Ph_2P — C_6H_4 — $PO(ONa)_2$		29
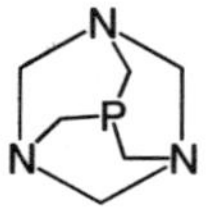		30
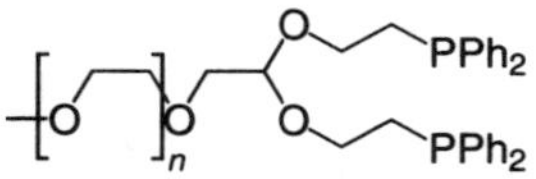	$n = 5,16$	31
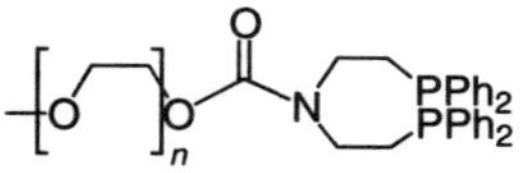	$n = 12,16,110$	9
$Ph_2PCH_2CH_2(OCH_2CH_2)_nPPh_2$	$n = 1,3$	32

TABLE 5.2 Chiral Water-Soluble Phosphine Ligands

Ligand (Abbreviation)	Ref.
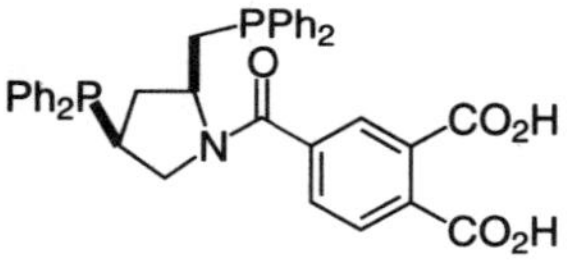	33
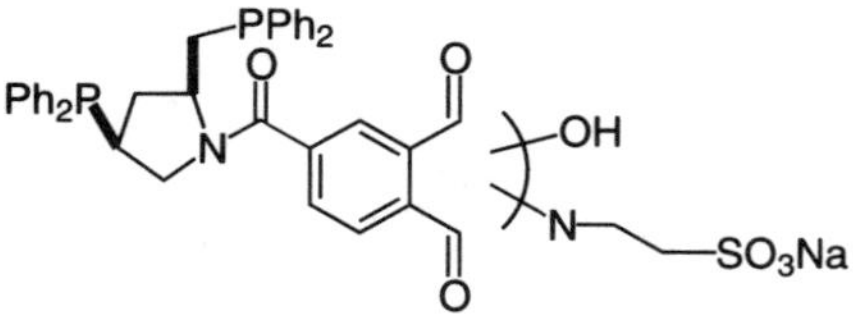	31
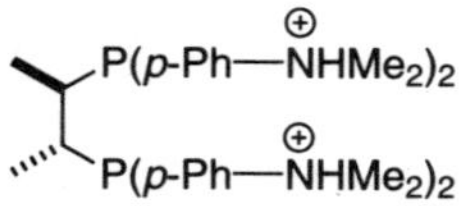	34
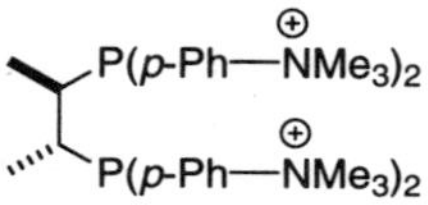	34

TABLE 5.2 Chiral Water-Soluble Phosphine Ligands (*continued*)

Ligand (Abbreviation)	Ref.
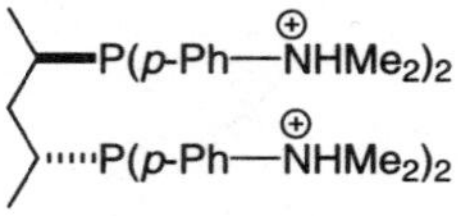 P(p-Ph—$\overset{\oplus}{N}HMe_2$)$_2$ P(p-Ph—$\overset{\oplus}{N}HMe_2$)$_2$	34
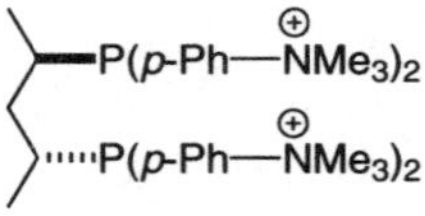 P(p-Ph—$\overset{\oplus}{N}Me_3$)$_2$ P(p-Ph—$\overset{\oplus}{N}Me_3$)$_2$	34
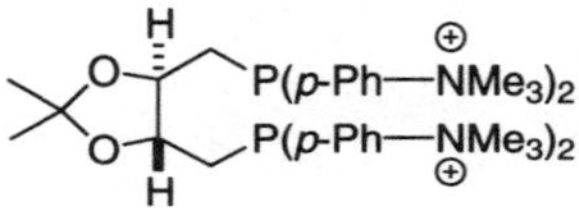 P(p-Ph—$\overset{\oplus}{N}Me_3$)$_2$ P(p-Ph—$\underset{\oplus}{N}Me_3$)$_2$	34
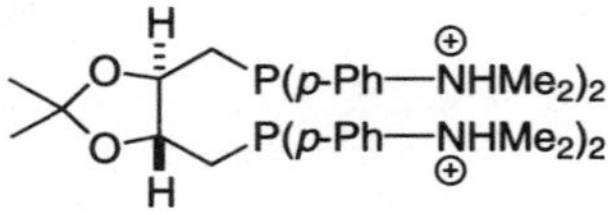 P(p-Ph—$\overset{\oplus}{N}HMe_2$)$_2$ P(p-Ph—$\underset{\oplus}{N}HMe_2$)$_2$	34
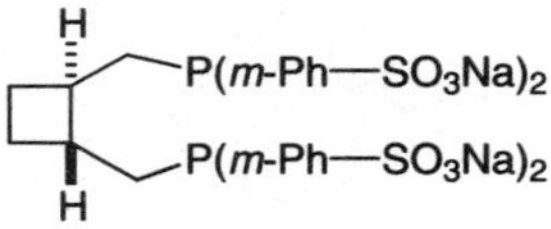 P(m-Ph—SO$_3$Na)$_2$ P(m-Ph—SO$_3$Na)$_2$	35
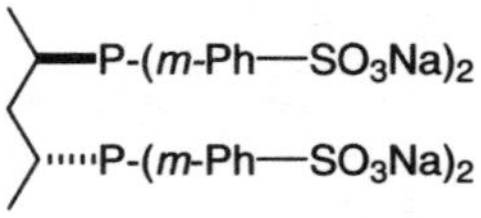 P-(m-Ph—SO$_3$Na)$_2$ P-(m-Ph—SO$_3$Na)$_2$ (bdppts)	35
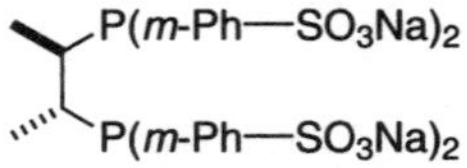 P(m-Ph—SO$_3$Na)$_2$ P(m-Ph—SO$_3$Na)$_2$	35
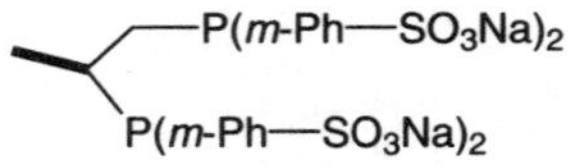 P(m-Ph—SO$_3$Na)$_2$ P(m-Ph—SO$_3$Na)$_2$	35
P(m —PhSO$_3$Na)$_2$	17

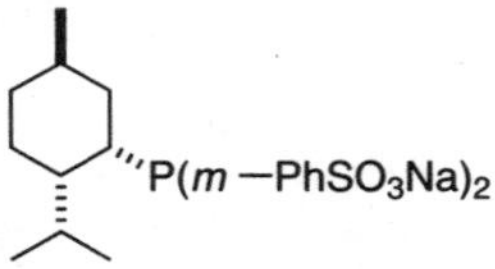

TABLE 5.2 Chiral Water-Soluble Phosphine Ligands (*continued*)

Ligand (Abbreviation)		Ref.
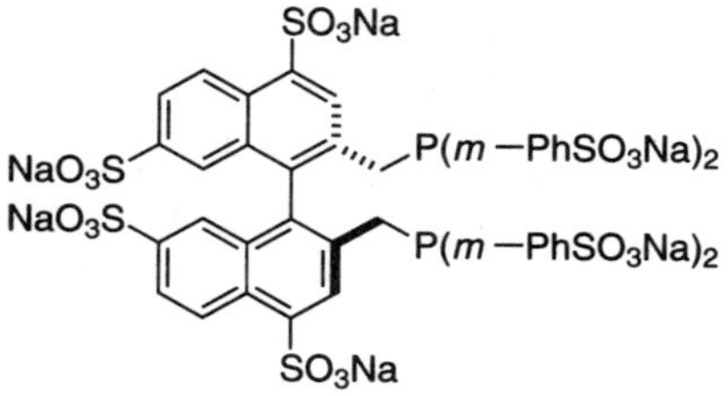	(BINAS-8)	14
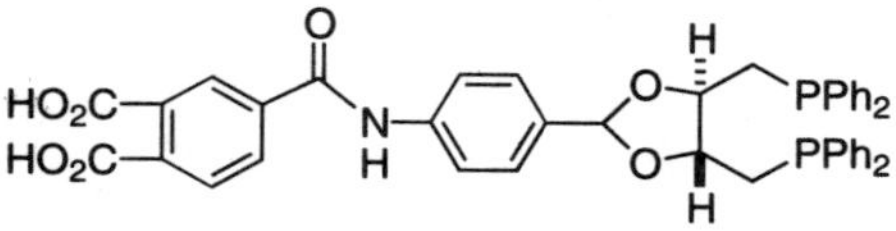		36
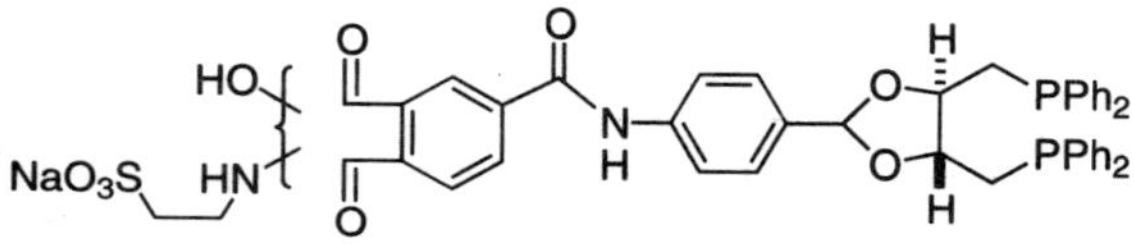		36
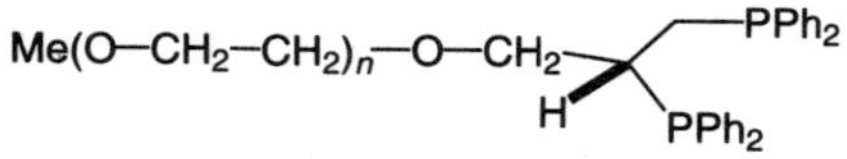	$n = 5, 18$	37
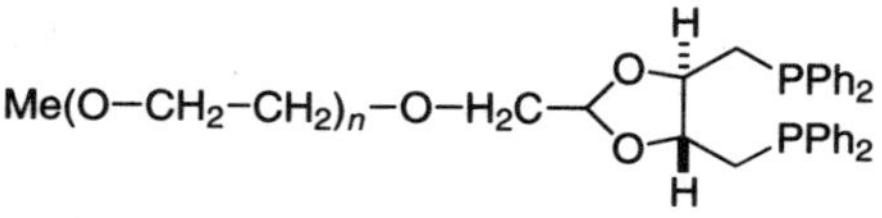	$n = 5, 16, 17$	37
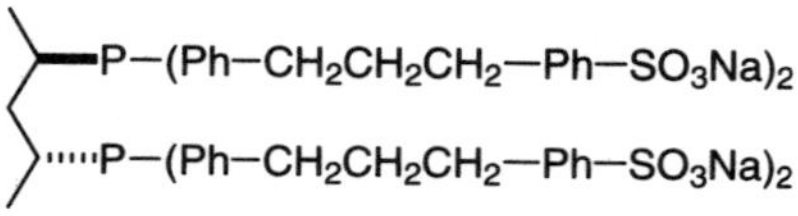		38
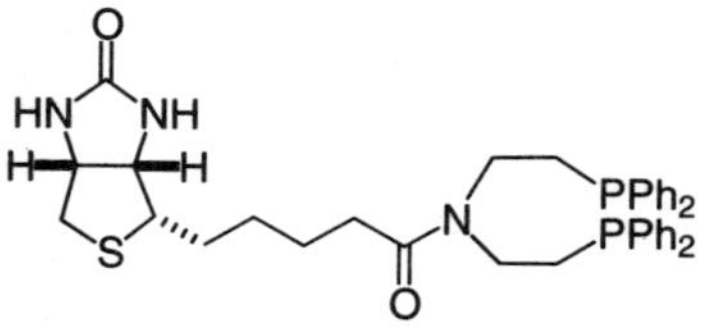		39
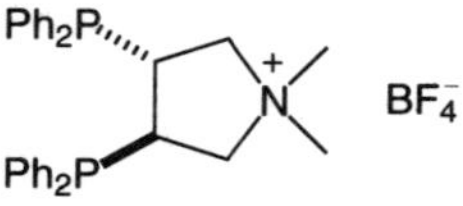		40
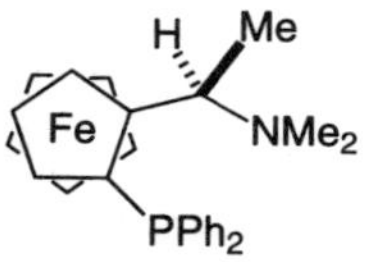		41

TABLE 5.2 Chiral Water-Soluble Phosphine Ligands (*continued*)

Ligand (Abbreviation)	Ref.
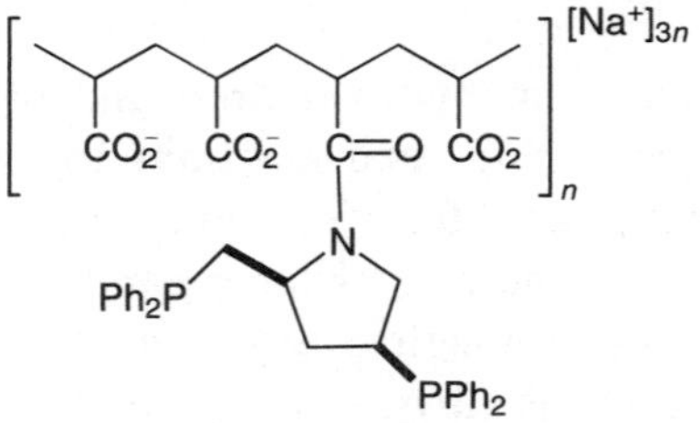	42

TABLE 5.3 Other Water-Soluble Ligands

Ligand (Abbreviation)	Ref.
$As(m\text{-}PhSO_3Na)_3$	43
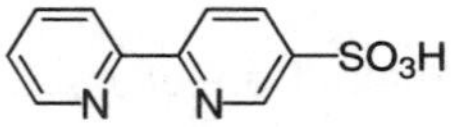	44
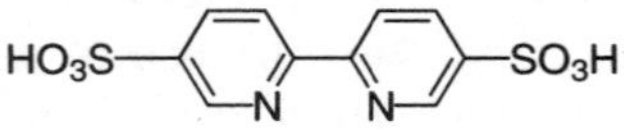	45
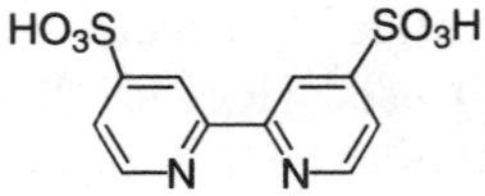	45
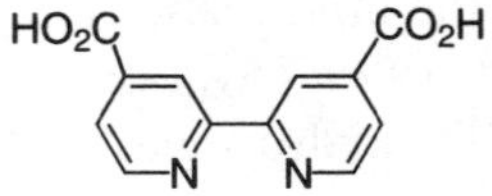	46
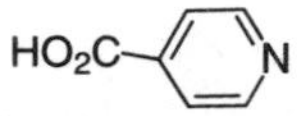	47
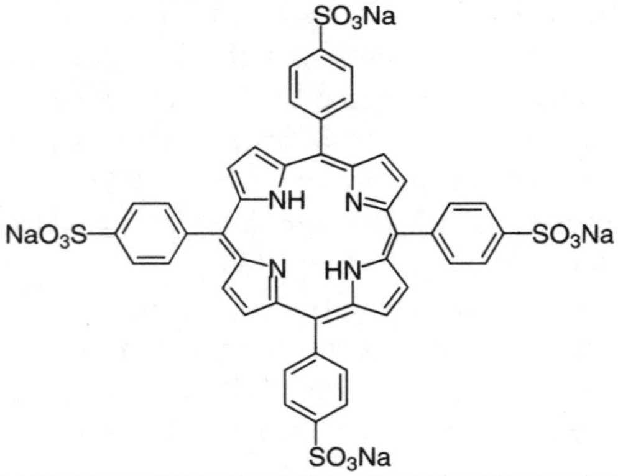	48

5.2 HYDROGENATION

5.2.1 General Hydrogenation

Hydrogenation, together with hydrocyanation, hydrosilation, and hydrostannation, is useful in transition-metal-catalyzed reactions in organic synthesis (49). Among the complexes used for hydrogenation, the Wilkinson catalyst, $RhCl(PPh_3)_3$, has been the most widely employed since it was first described in 1965 (50). Although hydrogenation involving aqueous media had been studied in the early 1960s in the presence of Co(II) salts and excess CN^- (51), the first attempt to carry out hydrogenation in aqueous solution with transition-metal (Rh, Pd, Pt) phosphine complexes came 10 years later and failed to give the corresponding product (52). The first successful catalytic hydrogenation in aqueous solution using a catalyst having water-soluble phosphine ligands was reported in 1975 (53). Water-soluble substrates were hydrogenated directly.

Dror and Manassen described the hydrogenation of cyclohexene in a two-phase system with $[RhCl_3 \cdot 3H_2O]$ in the presence of an excess of Ph_2PPhSO_3Na (TPPMS), in connection with the use of a cosolvent (54). The cosolvent has an effect on the reactivity of the catalyst, which increases in the order dimethylacetamide < dimethoxyethane < ethanol < methanol. Borowski et al. (55) used $[RhCl(TPPMS)_3]$ and $[RuHCl(TPPMS)_3]$ for hydrogenation without the use of a cosolvent. Some olefin migration was observed. On the other hand, Larpent et al. used the more water-soluble complex $[RhCl(TPPTS)_3]$, prepared *in situ* by mixing $[RhCl_3 \cdot 3H_2O]$ and the ligand, for hydrogenation (56). Many olefins were hydrogenated with 100% conversion and complete selectivity on the C—C double bond. Various functional groups survive the reaction condition. Through ^{31}P NMR study, the active catalyst was found to involve the phosphine oxide $OP(PhSO_3Na)_3$. The observation has been confirmed by further experiments in which no hydrogenation was observed when the amount of phosphine oxide was not sufficient (57). However, the hydrogenation of α,β-unsaturated aldehydes to saturated aldehydes proceeds without the involvement of TPPTS oxide (58). A rhodium amphos complex, $[Rh(NBD)(amphos)^2]^{3+}$, was found to be more air-stable and can be recycled easily (59).

A synthetically useful observation is the course of selectivity change when going from organic solvent to water for the hydrogenation of the diene **1** [Eq. (5.1)] (60). This unusual selectivity could be attributed to the coordination between the carboxylic group and the metal center.

$$
\text{1} \quad \diagup\!\!\!\diagdown\!\!\!\diagup\!\!\!\diagdown\!\!\!\diagup\!\!\!\diagdown\text{COOH} \xrightarrow[\text{RhCl[P}(p\text{-tolyl})_3]_3]{\text{H}_2} \left\{ \begin{array}{l} \diagup\!\!\!\diagdown\!\!\!\diagup\!\!\!\diagdown\!\!\!\diagup\!\!\!\diagdown\text{COOH} \quad \text{PhH, 66\%} \\[2em] \diagup\!\!\!\diagdown\!\!\!\diagup\!\!\!\diagdown\!\!\!\diagup\!\!\!\diagdown\text{COOH} \quad \text{PhH/H}_2\text{O, 81\%} \end{array} \right. \tag{5.1}
$$

The hydrogenation of alkenes and alkynes in water can also use silanes as hydrogen sources. Tour reported that by using palladium acetate as catalyst, triethoxysilane reduced C—C unsaturated bonds to saturation (61). For the reduction of alkynes, by carefully controlling the conditions, the reaction can be stopped at the alkene stage.

In addition to hydrogenation of C—C unsaturated bonds, hydrogenation also occurs on C—O double bonds in aqueous medium. Ruthenium-based complexes seem to be more effective for such hydrogenations. Joo and Benyei have shown that by using $RuCl_2(TPPMS)_2$ and sodium formate as hydrogen donor, a variety of aromatic and α,β-unsaturated aldehydes were transformed to the corresponding saturated alcohols in aqueous solution (62). Keto acids were hydrogenated to hydroxyl acids with $[RuH(OAc)(TPPMS)_3]$ and $[RuHCl(TPPMS)_3]$ (63):

$$
\overset{O}{\diagup\!\!\!\diagdown\!\!\!\diagup}\text{CO}_2\text{H} \xrightarrow[\text{H}_2,\ 60°\text{C, pH 1}]{[\text{HRuCl(TPPMS)}_3]} \overset{OH}{\diagup\!\!\!\diagdown\!\!\!\diagup}\text{CO}_2\text{H} \tag{5.2}
$$

By using the more water-soluble ligand, TPPTS, Grosselin et al. converted several unsaturated aldehydes into the corresponding unsaturated alcohols, with a 99% chemoselectivity on the carbonyl group [Eq. (5.3)] (64). Basset et al. found that the addition of NaI, which would assist the rapid formation of a metal–carbon bond, enhanced the reactivity (65).

$$
\diagup\!\!\!\diagdown\!\!\!\diagdown_{O} \xrightarrow[\text{H}_2]{[\text{HRuCl(TPPTS)}_3]} \underset{99\%}{\diagup\!\!\!\diagdown\!\!\!\diagdown\text{OH}} \tag{5.3}
$$

5.2.2 Asymmetric Hydrogenation

With the use of chiral phosphine ligands, hydrogenation of prochiral alkenes can provide optically active products. Asymmetric hydrogenation in aqueous media with several water-soluble chiral catalysts has been intensively investigated with a variety of substrates. For example, by using a water-soluble Rh catalyst, an α-acetamidoacrylic ester can be hydrogenated with up to 94% ee (enantiomeric excess) [Eq. (5.4)]

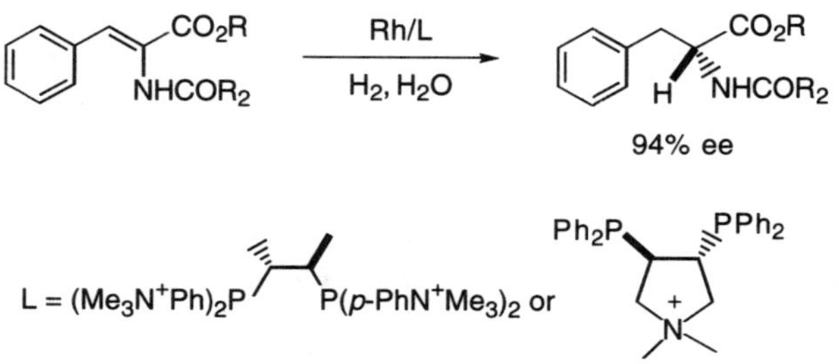
CO2R
NHCOR2
Rh/L
H2, H2O
CO2R
H
NHCOR2
94% ee
Ph2P
PPh2
L = (Me3N+Ph)2P
P(p-PhN+Me3)2 or
+
N

(5.4)

(40, 66). The hydrogenation of a prochiral imine gives a product with up to 96% ee [Eq. (5.5)] (67).

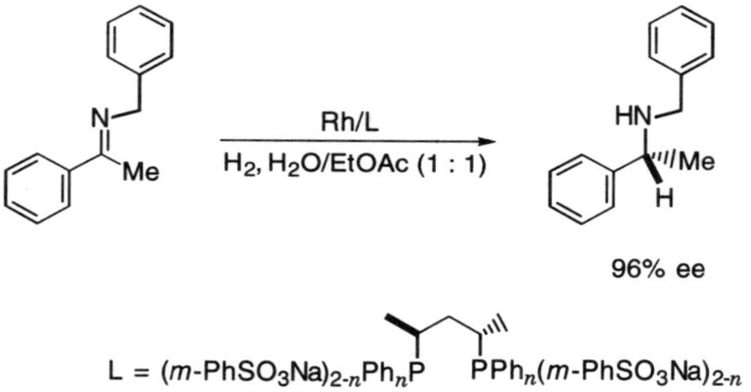
N
Me
Rh/L
H2, H2O/EtOAc (1 : 1)
HN
Me
H
96% ee
L = (m-PhSO3Na)2-nPhnP
PPhn(m-PhSO3Na)2-n

(5.5)

However, the optical yield of the product for asymmetric hydrogenation in aqueous media is generally lower than those obtained in organic solvents (68). Often, an increase in water content reduces the enantioselectivity. The decrease of enantioselectivity caused by addition of water was explained by Amrani and Sinou (37) through the use of Halpern's model (69) for asymmetric hydrogenation (Scheme 5.1). The two enantiomers are generated from two different pathways, and the optical yield is determined by the transition-state energy difference. The lower %ee in water was attributed to the smaller energy difference between the transition states.

By using a mixture of ethyl acetate and D_2O as solvent for hydrogenation, up to 75% deuterium is incorporated in the reduced product (70). This result indicates that the role of water here is not only as a sol-

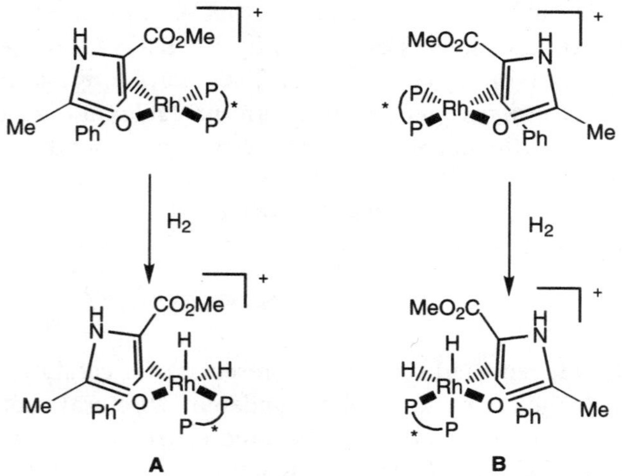

Scheme 5.1

vent. Research on asymmetric hydrogenation in aqueous media is still being actively pursued.

5.3 ALKENE ISOMERIZATION

In the presence of transition-metal complexes, organic compounds that are unsaturated or strained often rearrange themselves. One synthetically useful transition-metal-catalyzed isomerization is the olefin migration reaction. Two general mechanisms have been proposed for olefin migrations, depending on the type of catalyst employed (**A** and **B**) (Scheme 5.2) (71).

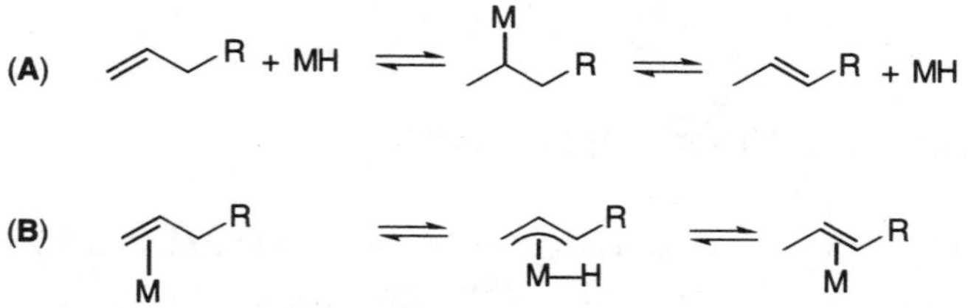

Scheme 5.2

Recently, Grubbs (72) demonstrated that olefin isomerization of allylic ethers and alcohols is catalyzed by $Ru(II)(H_2O)_6(tos)_2$ (tos = p-toluenesulfonate) in aqueous media. The olefin migration products, enols and enol ethers, thus generated are unstable and are hydrolyzed instantly to yield the corresponding carbonyl compounds:

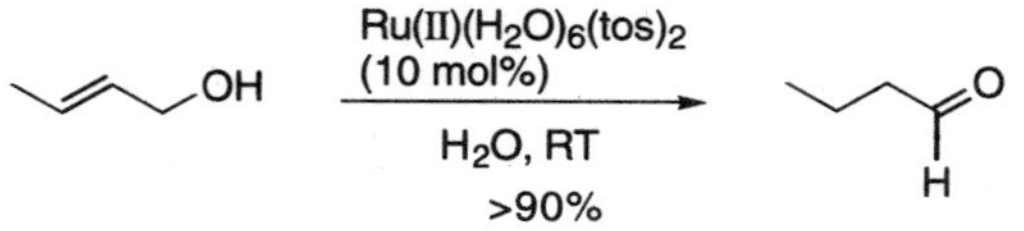

$$\text{(5.6)}$$

Li et al. (73) reported that in the presence of a catalytic amount of $RuCl_2(PPh_3)_3$, homoallylic alcohols undergo structural reorganization in which both the hydroxyl group and the olefin have been reshuffled in water [Eqs. (5.7) and (5.8)]. The reaction can be conceived of as an

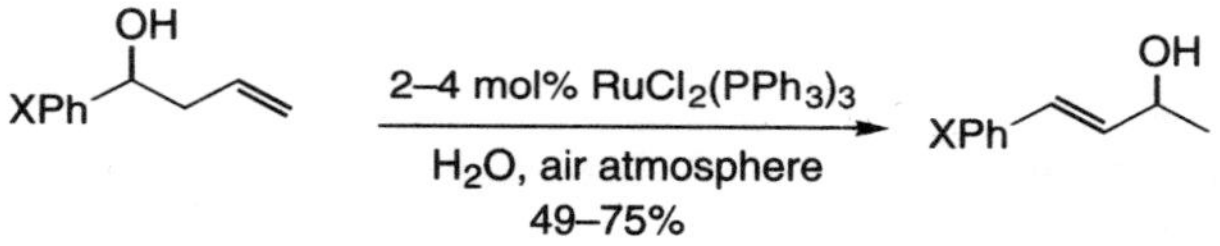

$$\text{(5.7)}$$

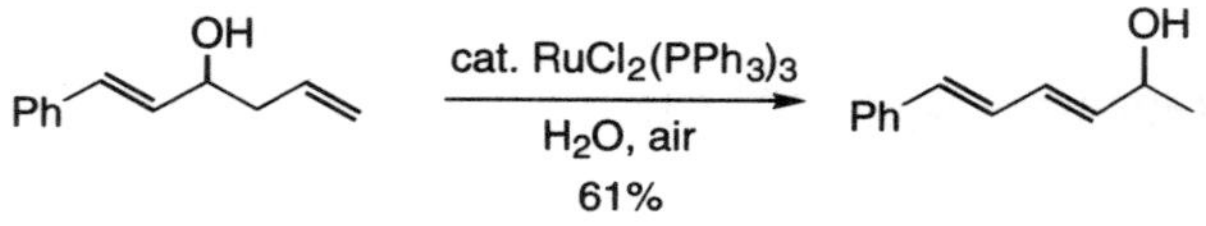

$$\text{(5.8)}$$

olefin migration followed by an allylic rearrangement. Thus, allyl alcohols are rearranged similarly. The use of water is critical to the success of this reaction [Eq. (5.9)].

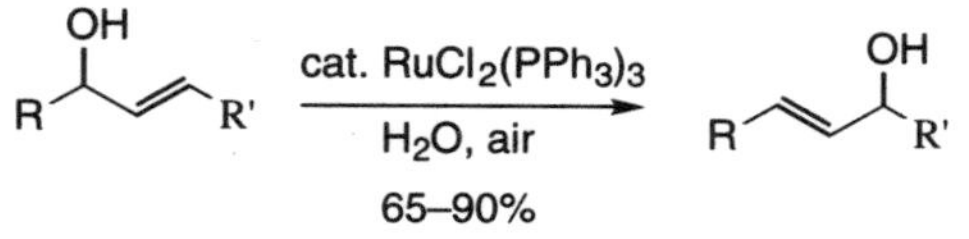

$$\text{(5.9)}$$

5.4 CARBONYLATION REACTIONS

The carbonylation of various organic compounds induced by transition metals is synthetically useful for the preparation of various carbonyl compounds (74). A key step for the transition-metal-catalyzed carbonylation reactions is the migration of an alkyl group to a CO that is coordinated to the catalyst. The overall outcome of this migration is the insertion of CO into the metal–alkyl bond of the reactive intermediate [Eq. (5.10)]. Depending on the use of different substrates and catalysts, there are different ways to form the alkyl–metal bond intermediate.

$$
\begin{array}{c}
\underset{\displaystyle R-ML_n}{\overset{\displaystyle \overset{O}{\underset{||}{C}}}{}}
\quad \underset{\displaystyle -CO}{\overset{\displaystyle CO}{\rightleftharpoons}} \quad
\underset{\displaystyle OC-ML_n}{\overset{\displaystyle R-C \diagup\!\!\!\!^{O}}{}}
\end{array}
\tag{5.10}
$$

5.4.1 Hydroformylation of Olefins

Hydroformylation is a major industrial process that produces aldehydes and alcohols from olefins, carbon monoxide, and hydrogen [Eq. (5.11)] (75). The reaction was discovered in 1938 by Roelen (76), who detected the formation of aldehydes in the presence of a cobalt-based catalyst.

$$
RCH{=}CH_2 + CO + H_2 \xrightarrow{\text{"Co"}}
$$
$$
RCH_2CH_2CHO + RCH(CH_3)CHO + \text{etc.}
\tag{5.11}
$$

In the 1950s, it was found that rhodium was much more active than cobalt, and the use of rhodium combined with phosphorous ligands proved to be even more effective. Since then, many improvements have been made on this process. However, all of them involve the tedious separation of the catalyst and products from the reaction system. Later improvements were based on the attachment of a normally soluble catalyst to an insoluble polymer support, in attempts to combine the virtues of both homogeneous and heterogeneous catalysts (77). Unfortunately, this approach encountered the problems of metal leaching into the solvent, lowered activity or selectivity, and facile oxidation of the ligands (78).

Recently, another approach to the separation of catalyst and product is based on the use of transition-metal complexes with water-soluble phosphine ligands and water as an immiscible solvent for the hydroformylation. Intensive research has been carried out in this regard. Most often, the catalyst is a rhodium complex. For example, by using the water-soluble complex $[HRh(CO)(Ph_2PPhSO_3Na)_3]$ (79), the hydroformylation can be carried out at a $>70°C$. However, with this monosulfonated ligand, some leaching of rhodium into the organic phase was observed (80).

In order to prevent the leaching of rhodium, the highly water-soluble tris-sulfonated ligand, $P(m\text{-}PhSO_3Na)_3$ (81, 82), was used. A variety of 1-alkenes were hydroformylated with this catalyst in high linear selectivity, generating the corresponding terminal aldehyde (83). More effective catalysts involving the use of other sulfonated phosphine ligands have also been reported (84).

The use of dinuclear rhodium complexes, $[Rh_2(\mu\text{-}SR)_2(CO)_2(P(m\text{-}PhSO_3Na)_3)_2]$, resulted in better performances of the catalyst for hydroformylation (85). Other water-soluble ligands, such as $Ph_2PCH_2CH_2NMe_3^+$ (86), p-carboxylatophenylphosphine (87), and sulfoalkylated tris(2-pyridyl)phosphine (88), have been used and are effective for hydroformylation in aqueous media.

For long-chain olefins, the hydroformylation generally proceeds slowly and with low selectivity in two-phase systems, due to their poor solubility in water. Monflier et al. reported recently a conversion of up to 100% and a regioselectivity of $\leq 95\%$ for the Rh-catalyzed hydroformylation of dec-1-ene in water, free of organic solvent, in the presence of partially methylated β-cyclodextrins [Eq. (5.12)] (89). Prior to this, hydroformylation in the presence of unmodified cyclodextrins had been studied by Jackson, but the results were rather disappointing (90).

$$+ CO + H_2 \xrightarrow{\substack{[Rh(acac)(CO)_2] \\ P(m\text{-}C_6H_4SO_3Na)_3}}$$

$$(OR)_a \quad (OR)_b \quad a + b = 21 \tag{5.12}$$

These interesting results are attributed to the formation of an alkene/cyclodextrin inclusion complex as well as the solubility of the chemically modified cyclodextrin in both phases.

In another interesting area of hydroformylation, Davis developed the concept of supported aqueous phase (SAP) catalysis (91). A thin, aqueous film containing a water-soluble catalyst adheres to silica gel with a high surface area. The reaction occurs at the liquid–liquid interface. Through SAP catalysis, the hydroformylation of very hydrophobic alkenes, such as octene or dicyclopentadiene, is possible with the water-soluble catalyst $[HRh(CO)tppts)_3]$.

Other metal complexes containing Pd, Ru, Co, or Pt are also used occasionally. For example, Khan investigated the complex $[Ru(EDTA)]^-$ for the hydroformylation of allyl alcohol and hexene in water. For the latter, very high linearities were obtained for the C_7 aldehyde generated (98–100%) (92). The hydroformylation reaction can also be performed by using methyl formate instead of carbon monoxide and hydrogen (93).

5.4.2 Carbonylation of Allylic and Benzylic Halides

The transition-metal-catalyzed carbonylation of allylic and benzylic compounds offers a useful method for the synthesis of β,γ-unsaturated acids (94). The requirement of a high carbon monoxide pressure and the low yield of the products limited the usefulness of the method in organic synthesis (95). In 1977, it was found that the carbonylation of benzyl bromide and chloride could be carried out by stirring aqueous sodium hydroxide and an organic solvent using a phase-transfer agent and a cobalt catalyst [Eq. (5.13)] (96, 97). Under high pressure and temperature, even benzylic mercaptans reacted similarly to give esters [Eq. (5.14)] (98).

$$PhCH_2Cl + CO \quad \xrightarrow[\text{PTC}]{[Co(CO)_4]^-,\ aq\ NaOH} \quad PhCH_2CO_2Na \qquad (5.13)$$

$$ArCH_2SH + CO + R'OH \quad \xrightarrow[\substack{850-900\ psi\ (lb/in.^2),\\ 190°\ C,\ 24\ h\\ 25-83\%}]{CO_2(CO)_8,\ H_2O} \quad ArCH_2COOR' + H_2S \qquad (5.14)$$

In the presence of a nickel catalyst, similar carbonylations of allyl bromide and chloride in aqueous NaOH were carried out at atmospheric pressure (99). The base concentration significantly influenced the yield and the product distribution. More recently, it was found that the palladium-catalyzed carbonylation of allyl chloride proceeded smoothly in a two-phase aqueous NaOH/benzene medium under atmospheric pressure at room temperature [Eq. (5.15)] (100). Catalysts with or without phosphorus ligands gave similar results, and the presence of hydroxide was essential. The reaction seemed to occur at the liquid–

liquid interface, because no phase-transfer agent was used. However, the addition of surfactants, such as $n\text{-}C_7H_{15}SO_3Na$ or $n\text{-}C_7H_{15}CO_2Na$, does accelerate the reactions (101):

$$CH_2{=}CHCH_2Cl + CO + ROH \xrightarrow[19-90\%]{[Pd]} CH_2{=}CHCH_2COOR \qquad (5.15)$$

$$R = H, CH_3, C_2H_5, \text{etc.}$$

By using a water-soluble palladium catalyst, 5-hydroxymethylfurfural is selectively carbonylated to the corresponding acid at 70°C, together with reduced product (102):

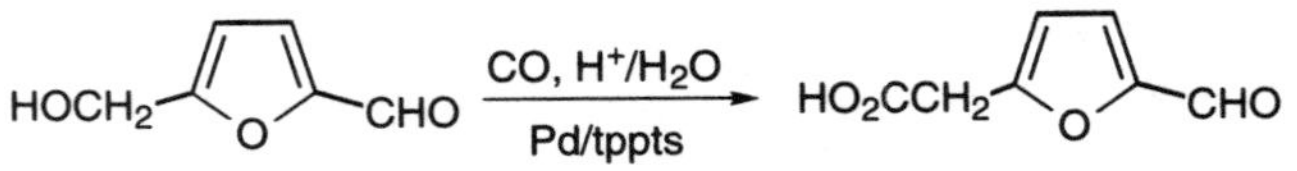

$$(5.16)$$

5.4.3 Carbonylation of Aryl Halides

The palladium-catalyzed carbonylation of aryl halides in the presence of various nucleophiles is a convenient method for synthesizing various aromatic carbonyl compounds (e.g., acids, esters, amides, thioesters, aldehydes, and ketones). Aromatic acids bearing different aromatic fragments and having various substituents on the benzene ring have been prepared from aryl iodides at room temperature under 1 atm CO in a mixed solvent of H_2O/DMF (1/1 or 1/2, v/v), and even in water alone, depending on the solubility of the substrate [Eq. (5.17)] (103).

$$Ar{-}I \xrightarrow[-I^-]{CO, OH^-, \text{"Pd"}} ArCOO{-} \xrightarrow{H^+} ArCOOH \qquad (5.17)$$

$$Ar = XC_6H_4, \text{Naph, Heteryl} \qquad (30-100\%)$$

The palladium(II) complexes $Pd(OAc)_2$, K_2PdCl_4, $PdCl_2(PPh_3)_2$, and $Pd(NH_3)_4Cl_2$ are used as the precursors of the catalyst, using either K_2CO_3 or NaOAc as the base. Iodoxyarenes can be carbonylated in water alone because of its solubility in the solvent (104). Recent work was done on the use of water-soluble catalysts (105).

Under the appropriate conditions of pressure and temperature, aryl mercaptans (thiophenols) can also be carbonylated in aqueous media with cobalt carbonyl as the catalyst (106).

5.4.4 Other Carbonylation Reactions

Transition-metal-catalyzed carbonylation of 1-perfluoroalkyl-substituted 2-iodoalkanes has been carried out in aqueous media to give carboxylic acids with a perfluoroalkyl substituent at the β position (107):

$$R_f\text{—}CH_2CHR'I + CO + H_2O \xrightarrow[\text{base} \quad 42\text{–}89\%]{\text{Pd, Co, or Rh cat.}} R_f\text{—}CH_2CHR'CO_2H \tag{5.18}$$

$$R_f = \text{perfluoroalkyl group}$$

Rhodium complexes also catalyze the carbonylation of both terminal and internal acetylenes (108). For example, a rhodium carbonyl cluster catalyzed the carbonylation of terminal acetylenes in water to give γ-lactones (109):

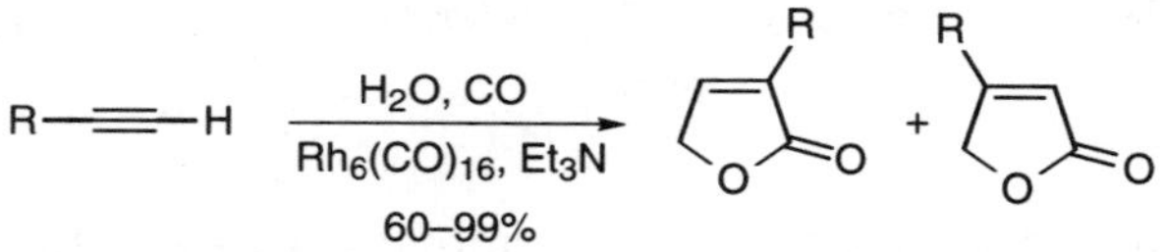

$$\tag{5.19}$$

Biscarbonylation in aqueous media has also been reported. The reaction of styrene oxide with carbon monoxide catalyzed by a cobalt complex in the presence of methyl iodide resulted in the incorporation of two molecules of carbon monoxide, giving the enol **2** (110):

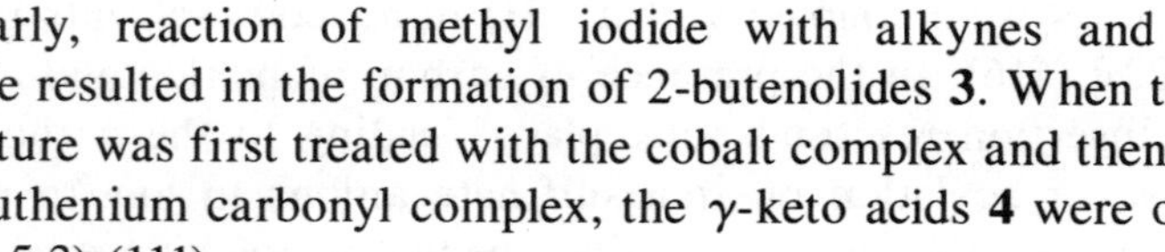

$$\tag{5.20}$$

Similarly, reaction of methyl iodide with alkynes and carbon monoxide resulted in the formation of 2-butenolides **3**. When the reaction mixture was first treated with the cobalt complex and then reacted with a ruthenium carbonyl complex, the γ-keto acids **4** were obtained (Scheme 5.3) (111).

Biscarbonylation of the vinylic dibromide **5** by carbon monoxide in NaOH solution gave the unsaturated diacid **6** in the presence of a palladium catalyst [Eq. (5.21)] (112). By using palladium chloride as a catalyst in a water/THF mixture, biscarbonylation products were also

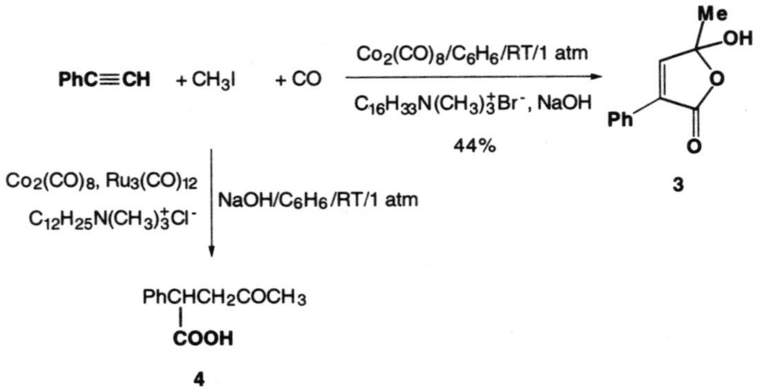

Scheme 5.3

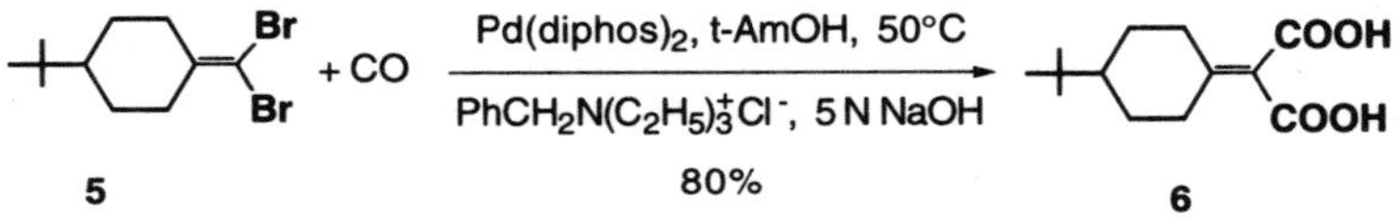

$$(5.21)$$

formed from the carbonylation of terminal alkynes with formic acid (113). Carbonylation of ammonia, diethylamine, and triethylamine can be catalyzed by $[Ru(EDTA-H)](CO)]^-$ in aqueous media to give amides (114). Low-molecular-weight alkanes are carbonylated under acidic conditions by oxygen and CO in water, catalyzed by palladium, platinum, or rhodium catalysts (115). In this way, acetic acid is generated from methane.

5.5 ALKYLATION AND COUPLING REACTIONS

Transition-metal-catalyzed coupling reactions have become increasingly important in C—C bond formation. As early as 1970, arylsulfinic acids were coupled with Pd(II) in aqueous solvents to biaryls [Eq. (5.22)] (116). In the presence of carbon monoxide and olefins or nitriles, insertion reactions took place, leading to the carbonylation, vinylation, or acylation of arenesulfinate anions in low to moderate yields. However, the reaction requires the use of stoichiometric amounts of palladium.

$$2 \text{ ArSO}_2\text{Na} + \text{Na}_2\text{PdCl}_4 \xrightarrow[1-36\%]{\text{H}_2\text{O}} \text{Ar-Ar} + 2 \text{ SO}_2 + \text{Pd} + 4 \text{ NaCl} \qquad (5.22)$$

Generally, two types of intermediates are involved in palladium- (and nickel-) catalyzed reactions: the π-allyl palladium (nickel) complexes and the oxidative addition intermediate of organic halides to Pd(0) [or Ni(0)] (Scheme 5.4). Both intermediates are subjected to the attack of reagents, generating useful products and reproducing the active catalyst.

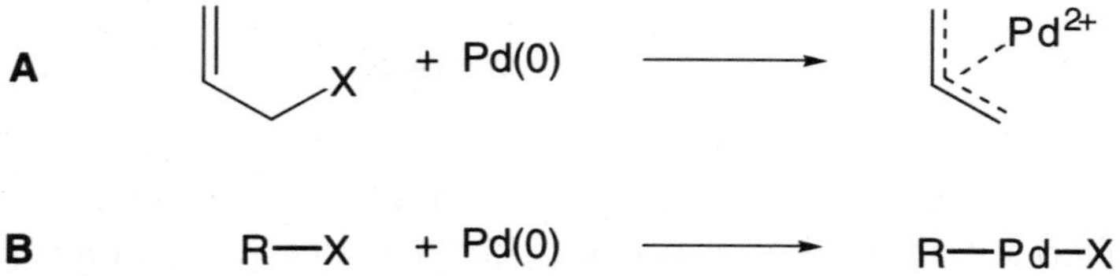

A

B R—X + Pd(0) ⟶ R—Pd—X

Scheme 5.4

In the first case, the leaving group can be halides, carboxylates, phosphates, or similar compounds. The π-allyl complexes generated are electrophilic and undergo reactions with a variety of nucleophiles, typically stabilized carbanions (117). Most importantly, the π-allyl complexes can be generated *in situ* using a catalytic amount of a Pd(0) compound. The reactive Pd(0) species is regenerated in an elimination step (Scheme 5.5).

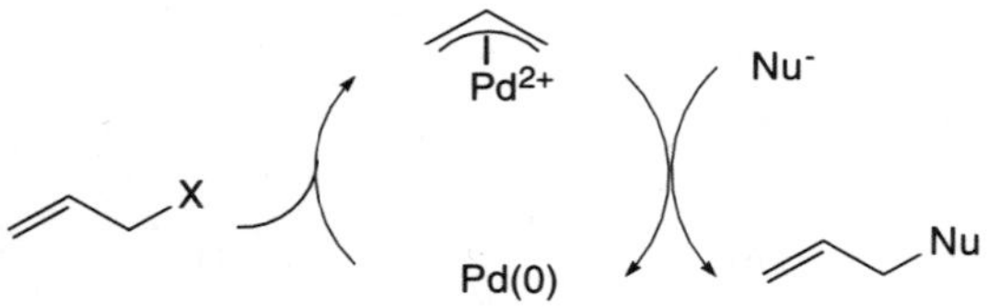

Scheme 5.5

Coupling reactions involving π-allyl palladium intermediates have recently been investigated in aqueous media (118). For example, with the use of a water-soluble palladium catalyst, allyl acetates couple with α-nitroacetate in water in the presence of triethylamine [Eq. (5.23)].

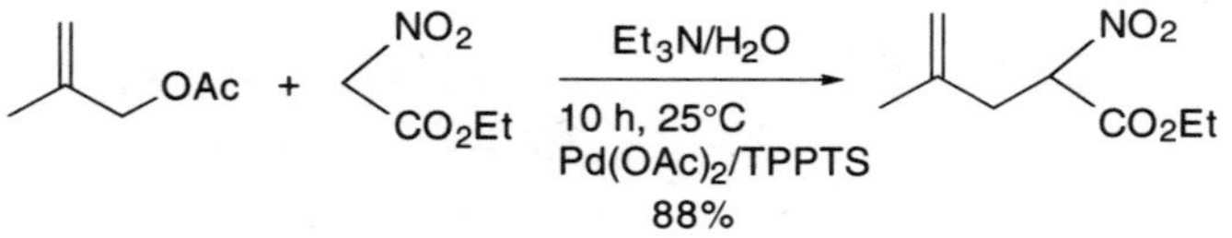

Other compounds bearing an active hydrogen can be used as well in place of α-nitroacetate [Eq. (5.24)].

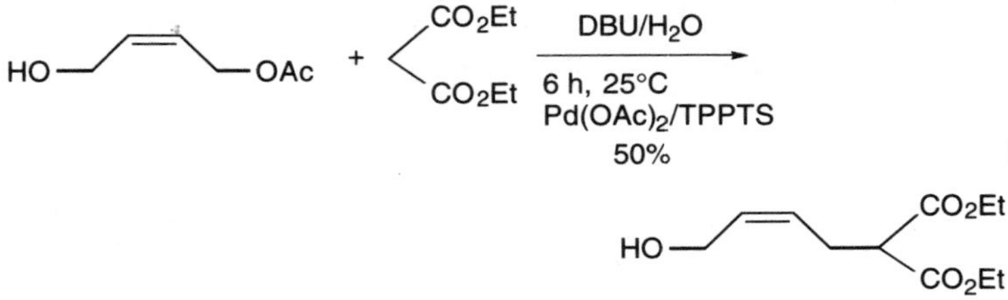

(5.24)

Heteroatom nucleophiles can also be used for this reaction. The reaction of an amine with allyl acetate generated the *N*-allylation product [Eq. (5.25)] (119). Palladium(0)-catalyzed reaction of uracil with cin-

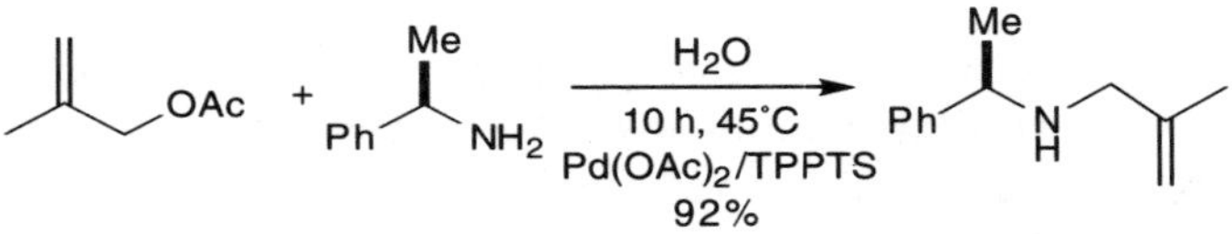

(5.25)

namyl acetate in aqueous acetonitrile gives 1-cinnamyluracil regioselectively in the presence of DBU [Eq. (5.26)]. On the other hand, the

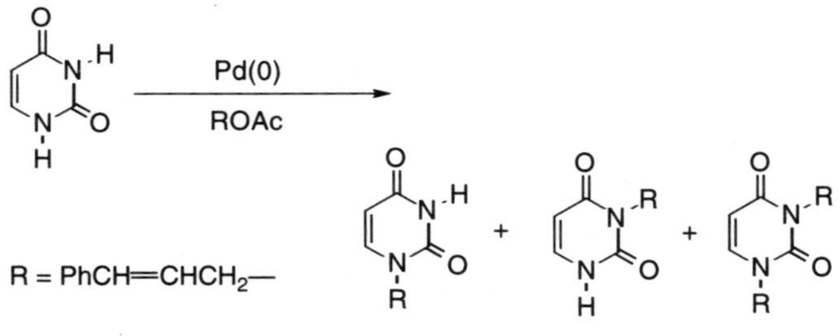

(5.26)

R = PhCH=CHCH$_2$—

in H$_2$O/MeCN (17 : 2)	80%	0	0
in DMSO	38	7	9

same reaction in DMSO led to all *N*-allylation products. Similar results were obtained with thymine. Nucleophiles such as azide [Eq. (5.27)],

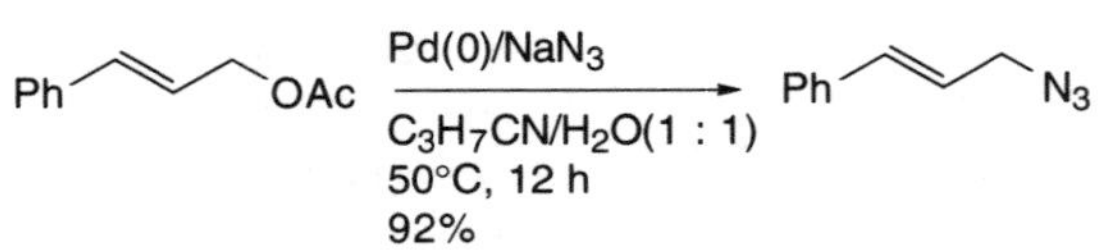

(5.27)

and toluenesulfinate [Eq. (5.28)] reacted similarly to give quantitative yields of the corresponding allyl azide and allylsulfone (120).

$$\text{Ph}\diagdown\diagup\diagdown\text{OAc} \xrightarrow[\substack{\text{C}_3\text{H}_7\text{CN/H}_2\text{O}\,(1:1)\\50°\text{C, 12 h}\\95\%}]{\text{Pd(0)/MePhSO}_2\text{Na}} \text{Ph}\diagdown\diagup\diagdown\text{SO}_2\text{PhMe} \qquad (5.28)$$

The reaction can also be used for the removal of allyloxycarbonyl-protected functional groups [Eq. (5.29)] (121). The method has been

$$\underset{\text{R—Z}}{\overset{\text{O}}{\diagdown\!\!\diagup}}\text{O}\diagdown\diagup \xrightarrow[70-99\%]{\substack{\text{Pd(OAc)}_2/\text{TPPTS}\\ \text{Et}_2\text{NH}}} \text{R—ZH} + \text{Nu}\diagdown\diagup\diagdown \qquad (5.29)$$

successfully used for deprotection of a wide range of secondary amines. Both homogeneous aqueous acetonitrile or biphasic diethyl ether/water system are suitable for the removal of the alloc moiety from nitrogen- and oxygen-based functional groups.

Examples involve the use of organomercury reagents as nucleophiles in aqueous media are also known. Bergstrom studied the synthesis of C_5-substituted pyrimidine nucleosides **8** in aqueous media via the mercurated intermediate **7** using Li_2PdCl_4 as a catalyst [Eq. (5.30)] (122).

$$(5.30)$$

7   **8**

Mertes investigated the coupling of the 5-mercuriuridines **9** with styrenes in aqueous media, resulting in alkylation of the uracil nucleotides [Eq. (5.31)] (123). Carbon alkylation of C_5 of uracil ring in the ribo- and deoxyribonucleosides and nucleotides was obtained in high yields by this method. The reaction was not affected adversely by the presence of phosphate groups or sugar hydroxyls, and was compatible with nitro, amino, and azido substitution on the phenyl ring of the styrene. A similar reaction was used by Langer et al. in the synthesis of 5-(3-

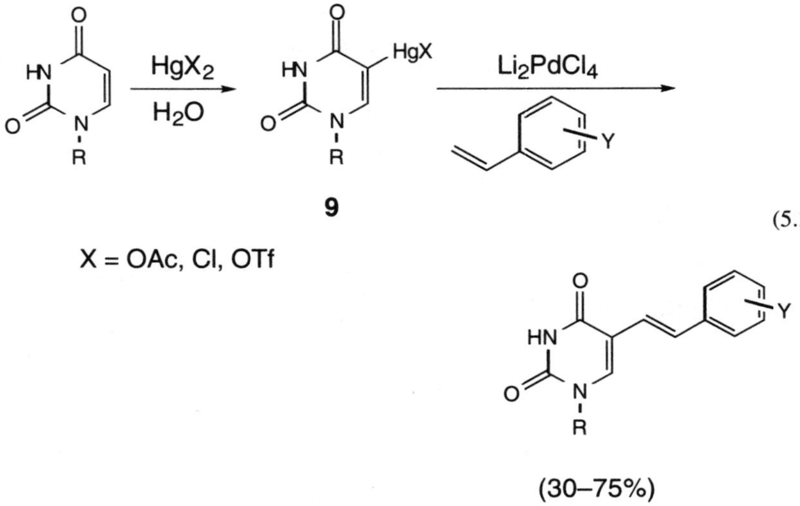

9

$$(5.31)$$

X = OAc, Cl, OTf

(30–75%)

amino)allyluridine and deoxyuridine-5′-triphosphates (AA-UTP and AA-dUTP) (**10**) (124).

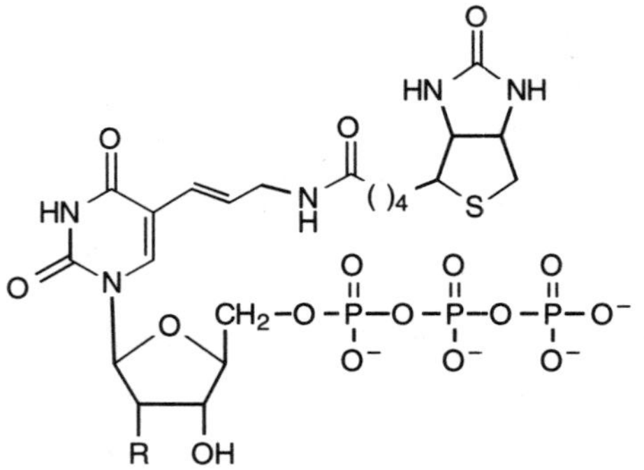

10 R = OH, H

The reaction between aryl (or alkenyl) halides and alkenes in the presence of a catalytic amount of a palladium compound to give substitution of the halides by the alkenyl group is commonly referred to as the *Heck reaction* (125). The catalytic process of the Heck reaction is depicted in Scheme 5.6.

Recently, both inter- and intramolecular Heck reactions have been performed in aqueous media. Palladium-catalyzed reactions of aryl halides with acrylic acid or acrylonitrile gave the corresponding cou-

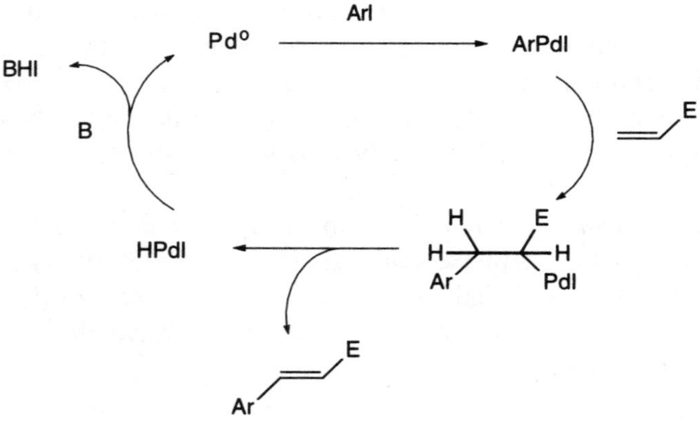

Scheme 5.6

pling products in high yields with a base (NaHCO$_3$ or K$_2$CO$_3$) in water [Eq. (5.32)] (126). The reaction provides a new and simple method for

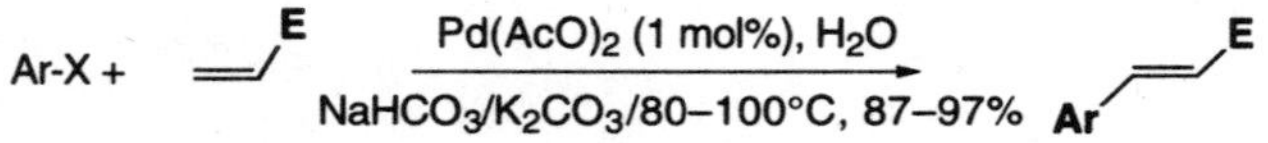

$$\text{Ar-X} + \quad \overset{E}{=\!\!/} \quad \xrightarrow[\text{NaHCO}_3/\text{K}_2\text{CO}_3/80-100°\text{C}, 87-97\%]{\text{Pd(AcO)}_2 \text{ (1 mol\%), H}_2\text{O}} \quad \underset{\text{Ar}}{\overset{E}{/\!\!=\!\!/}} \tag{5.32}$$

the synthesis of substituted cinnamic acids and cinnamonitriles. Recently, such reactions were carried out by using a water-soluble phosphine ligand (127). These reactions are usually complete within several hours at low temperatures (25–66°C). Iodobenzoic acid can be used directly for coupling with acrylic acid [Eq. (5.33)]. Diaryliodonium salts react similarly [Eq. (5.34)] (128).

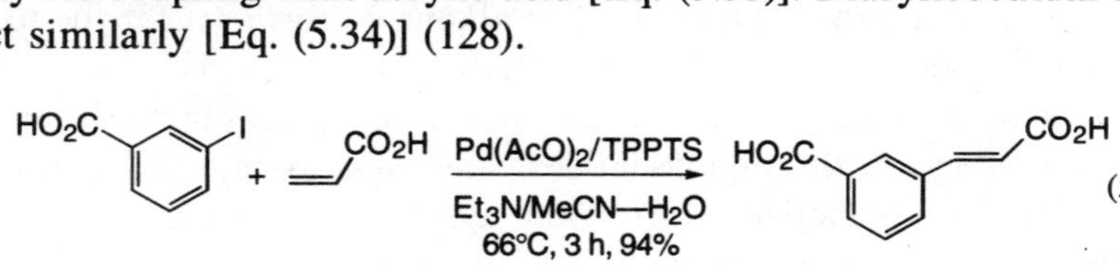

(5.33)

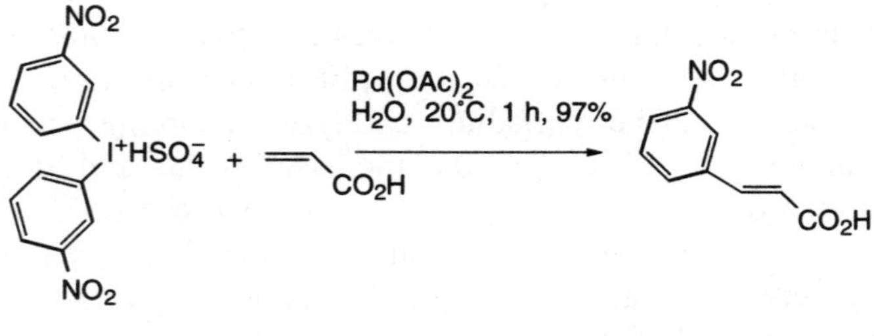

(5.34)

Jeffery studied the reaction under phase-transfer conditions and found that the presence of water is the determinant for the efficiency of quaternary ammonium salt in the palladium-catalyzed vinylation of organic halides using an alkaline metal carbonate as the base (129). The phase-transfer procedure can be performed without the organic co-solvent.

In aqueous DMF, the reaction can be applied to the formation of C—C bonds in a solid-phase synthesis [Eq. (5.35)] (130). The reaction proceeds smoothly and leads to moderate to high yields of product under mild conditions. The optimal conditions involve the use of 9:1 DMF—water mixture.

$$(5.35)$$

Recently, Parsons investigated the viability of the aqueous Heck reactions under superheated conditions (131). A series of aromatic halides were coupled with styrenes under these conditions. The reaction proceeded to approximately the same degree at 400°C as at 260°C. Some 1,2-substituted alkanes can be used as alkene equivalents for the high-temperature Heck-type reaction in water (132).

The general catalytic cycle for the cross-coupling reaction of organometallics with organohalides involves oxidative addition–transmetalation–reductive elimination sequences (Scheme 5.7), which is different from the cross-coupling of alkenes.

The coupling between alkenyl and arylhalides with organostannanes in the presence of a palladium catalyst is referred to as the *Stille reaction* (133). Although it was known that the addition of water to the organic reaction medium in the palladium-catalyzed coupling reaction of organostannanes with vinyl epoxides increased yields and affects the regio- and stereochemistry (134), the Stille coupling reaction in aqueous media was investigated only recently. Davis (135) reported a case of coupling between **11** and **12** in aqueous ethanol [Eq. (5.36)]. The reaction gave high yield of the coupled product, which hydrolyzed *in situ*.

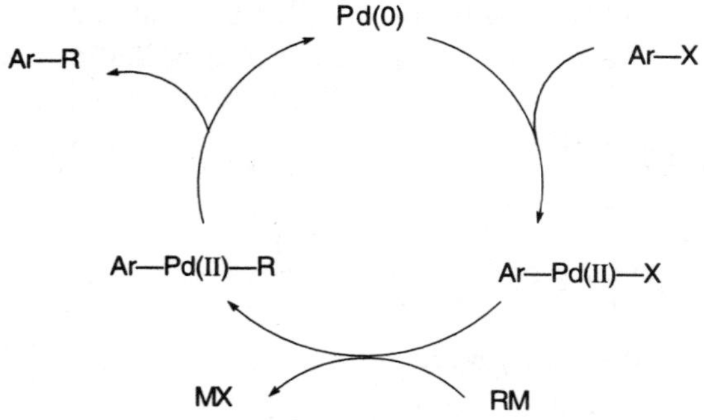

Scheme 5.7

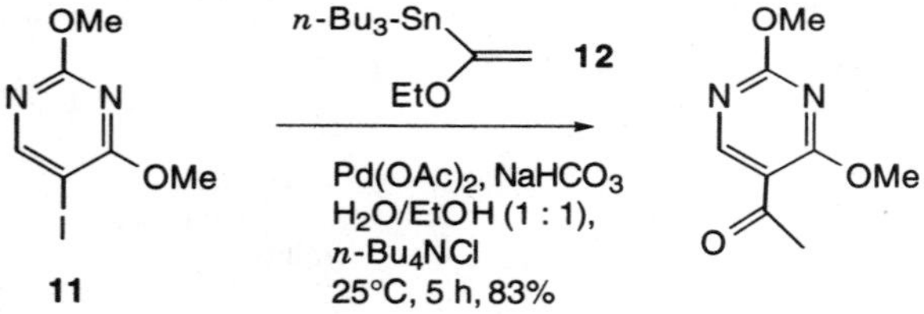

$$(5.36)$$

Collum (136) recently reported that while the Stille coupling can proceed without using a phosphine ligand, the addition of a water-soluble ligand improved the yield of the reaction. Water-soluble aryl and vinyl halides were coupled with alkyl-, aryl-, and vinyltrichlorostannane derivatives in this way:

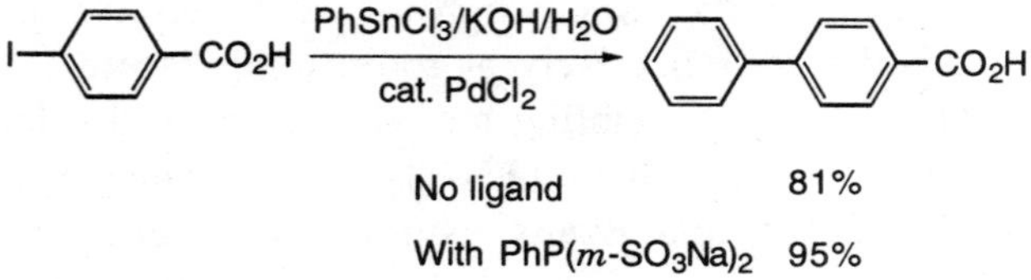

$$(5.37)$$

No ligand   81%

With PhP(m-SO$_3$Na)$_2$ 95%

Arenediazonium chlorides and hydrogen sulfates react with tetra-methyltin in aqueous acetonitrile in the presence of a catalytic amount of palladium acetate to give high yields of substituted toluenes (137):

$$ArN_2X + Me_4Sn \xrightarrow[\substack{20°C,\ 2\ h,\\30-88\%}]{\substack{Pd(OAc)_2\\H_2O-MeCN\ (1:1)}} ArMe$$

X = Cl or HSO$_4$

$$(5.38)$$

The cross-coupling reaction of alkenyl and aryl halides with organoborane derivatives in the presence of a palladium catalyst and a base, known as the *Suzuki reaction*, has often been carried out in an organic/aqueous mixed solvent [Eq. (5.39)] (138). Thus, dienes are con-

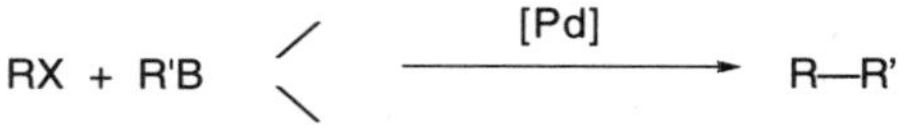

$$RX + R'B \quad \diagdown \quad \xrightarrow{\text{[Pd]}} \quad R\!-\!R'$$

(5.39)

R = 1-alkenyl or aryl; R' = aryl or alkyl

X = Br or I

veniently prepared from the corresponding alkenylborane and vinyl bromide in refluxing THF in the presence of Pd(PPh$_3$)$_4$ and an aqueous NaOH solution [Eq. (5.40)] (139).

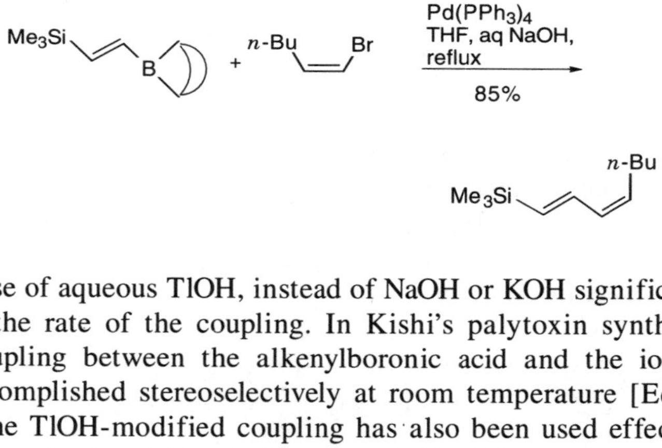

(5.40)

The use of aqueous TlOH, instead of NaOH or KOH significantly increased the rate of the coupling. In Kishi's palytoxin synthesis, the cross-coupling between the alkenylboronic acid and the iodoalkene were accomplished stereoselectively at room temperature [Eq. (5.41)] (140). The TlOH-modified coupling has also been used effectively in Roush's synthesis of kijanimicin (**13**) (141), Nicolaou's synthesis of (12*R*)-HETE (**14**) (142), and Evans's synthesis of rutamycin B (**15**) (143).

The preparation of biaryls by the Suzuki reaction was initially carried out in a mixture of benzene and aqueous Na$_2$CO$_3$ (144). However, the reaction proceeds more rapidly in a homogeneous medium, such as aqueous DME. This condition works satisfactorily in most aryl–aryl couplings (145).

Recently, Casalnuovo and Calabrese reported that using the water-soluble palladium(0) catalyst Pd[PPh$_2$(m-C$_6$H$_4$SO$_3$M)]$_3$ (M = Na$^+$, K$^+$), various aryl bromides and iodides reacted with aryl and vinyl boronic acids, terminal alkynes and dialkyl phosphites to give the

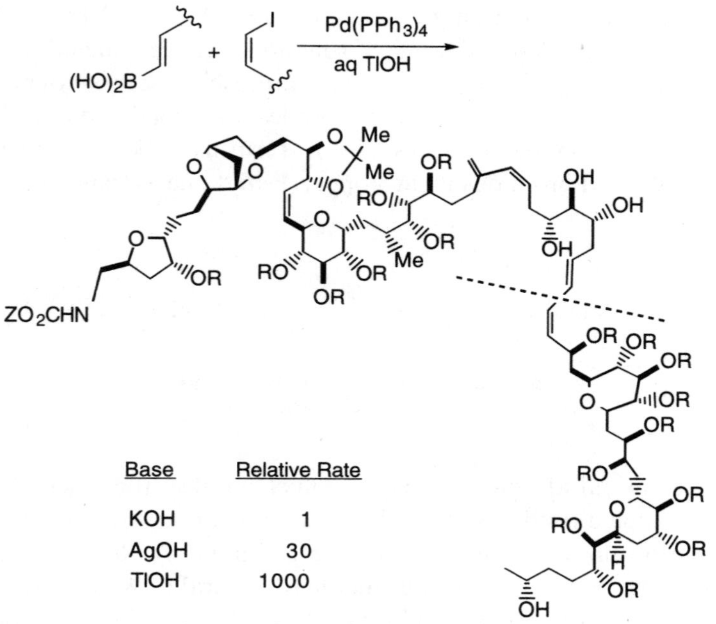

(5.41)

Base	Relative Rate
KOH	1
AgOH	30
TlOH	1000

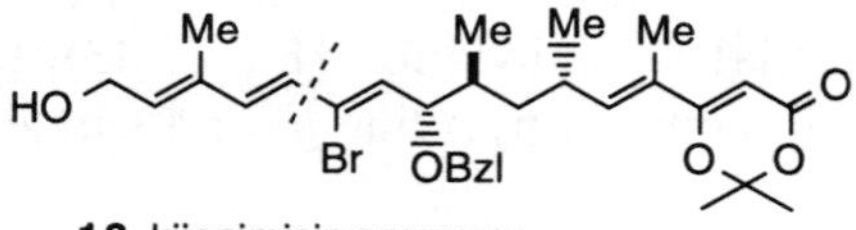

13 kijanimicin precusor

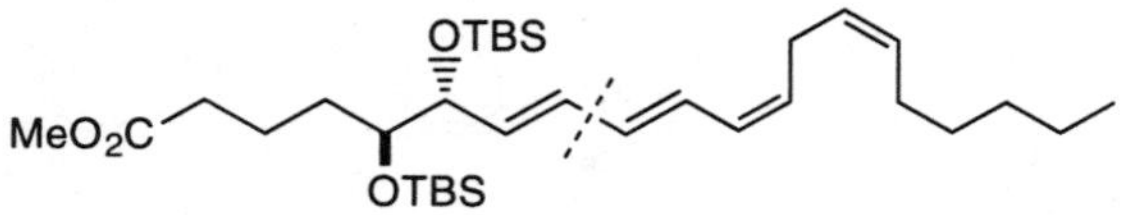

14 HETE

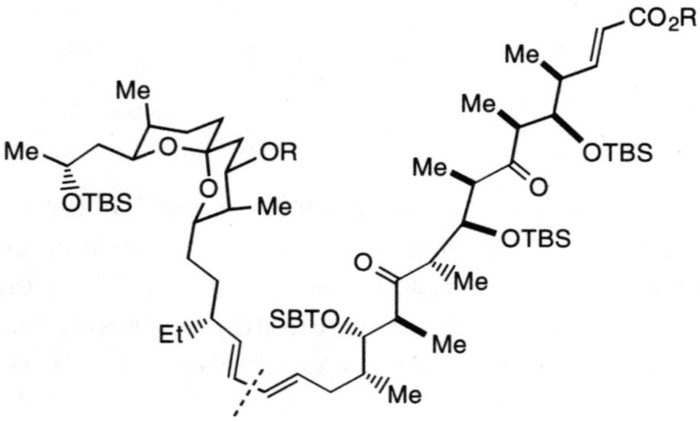

15 rutamycin precusor

cross-coupling products in high yields in water [Eq. (5.42)] (146). This reaction can tolerate a broad range of functional groups, including those present in unprotected nucleotides and aminoacids. Cross-coupling of boronic acids or esters with alkenyl iodides was conducted similarly, generating functionalized dienes (147). Terminal alkynes were also coupled with allyl bromides using copper(I) and phase-transfer catalyst (148).

$$R-X \ + \ R'-Y \quad \xrightarrow[47-100\%]{\text{[Pd]}} \quad R-R'$$

(5.42)

X = Br, I; Y = H, B(OR)$_2$; R = aryl, heteroaromatic;
R' = aryl, vinyl, alkynyl, P(O)(OR)$_2$

When terminal alkynes are used as nucleophiles, the reaction gives the cross-coupled products (149). The use of a copper salt as cocatalyst significantly improved the yield of the reaction (150). Recently, the reaction was investigated in aqueous media by several groups. In an initial investigation, Piskunov et al. observed that the reaction of 1-iodo-4-methylaminoanthraquinone with a propargyl alcohol in triethylamine was accompanied by partial demethylation [Eq. (5.43)] (151). In contrast, the same reaction in aqueous pyridine gave a single product.

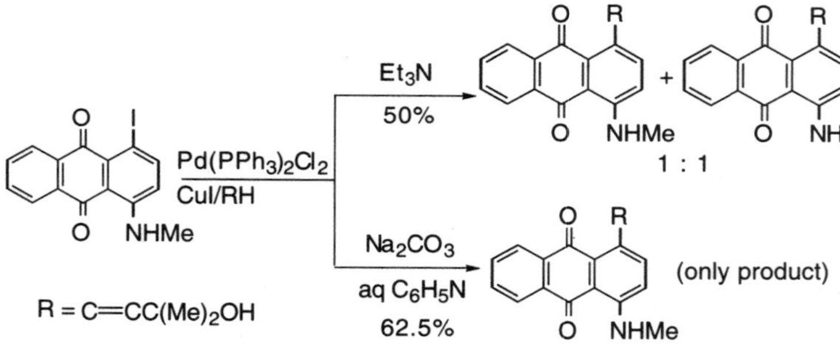

(5.43)

When a water-soluble palladium complex is used as the catalyst, unprotected nucleosides, nucleotides, and aminoacids undergo coupling with terminal acetylenes in aqueous acetonitrile (152). Compound T-505 (**16**), part of a family of chain-terminating nucleotide reagents used in DNA sequencing and labeling, was synthesized by this route in 50% yield:

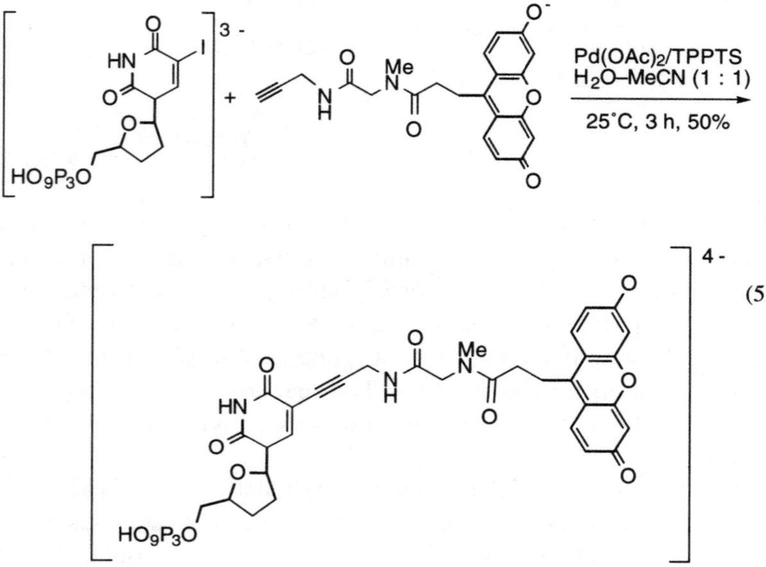

(5.44)

16 T-505

Genet (153) and Beletskaya (154) carried out more detailed studies of the aqueous reaction recently. A variety of aryl and vinyl iodides were coupled with terminal acetylenes in aqueous media with either a water-soluble catalyst or a non-water-soluble catalyst. The reaction can also be carried out without the use of a phosphine ligand in a water-alcohol emulsion in the presence of cetyltrimethylammonium bromide (CTAB) as an emulsifier (155). Diynes were prepared similarly by reaction of alkynyl bromide with terminal alkynes (156).

Pd(OAc)$_2$/TPPTS has been used as an efficient catalyst in a sequential two-step reaction in aqueous media for the coupling of 2-iodoaniline or 2-iodophenol with terminal alkynes to give the corresponding indoles or benzofurans in good yield (157):

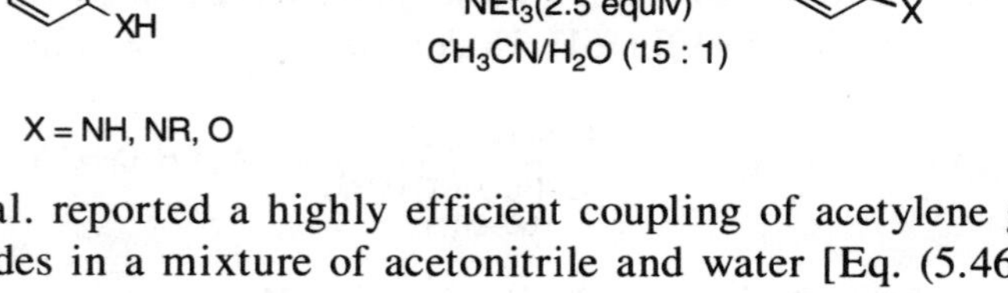

(5.45)

X = NH, NR, O

Li et al. reported a highly efficient coupling of acetylene gas with aryl halides in a mixture of acetonitrile and water [Eq. (5.46)] (158).

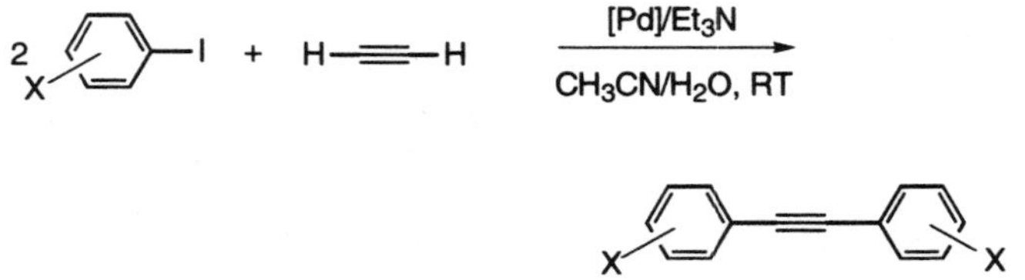

$$(5.46)$$

The conditions are generally milder, and the yields are better than previously reported results in organic solvents. A variety of aromatic halides are coupled to give the corresponding bisarylacetylenes. Both a water-soluble palladium catalyst [generated *in situ* from Pd(OAc)$_2$/TPPTS] and water-insoluble catalyst [generated *in situ* from Pd(OAc)$_2$/PPh$_3$] can be used for the reaction. The reaction can be carried out in water alone. The reaction proceeds equally well with or without CuI as a cocatalyst.

Biaryls were obtained in good yields by reacting diphenyldifluorosilane or diphenyldiethoxysilane with aryl halides in aqueous DMF at 120°C in the presence of KF and a catalytic amount of PdCl$_2$ [Eq. (5.47)] (159). Phosphine ligands are not required for the reaction.

$$ArBr + Ph_2SiF_2 \xrightarrow[\substack{DMF-H_2O,\ 120°C \\ KF}]{2\ mol\%\ PdCl_2} Ar-Ph + [PhSiF_3] \qquad (5.47)$$

In the presence of hydrogen gas, aryl halides homocoupled to give biaryl compounds in an aqueous/organic microemulsion [Eq. (5.48)] (160). However the yield is only moderate, typically 30–50%.

$$2\ ArX \xrightarrow[\substack{H_2O-BuOH}]{H_2,\ cat.\ PdCl_2,\ K_2CO_3} Ar-Ar \qquad (5.48)$$

X = I, Br
Ar = Ph, *p*-MePh, *p*-MeOPh, *p*-ClPh, *m*-CF$_3$Ph

The nucleophiles for these reactions can also be noncarbon nucleophiles. Thus, diphenylamine was coupled with aryl iodides in an alkaline aqueous–ethanol emulsion in the presence of a phase-transfer catalyst (CTAB) [Eq. (5.49)] (161). Under the same conditions, reaction

$$Ar_2NH + Y-C_6H_4-I \xrightarrow[\substack{H_2O-BuOH,\ CTMAB \\ 80-90\%}]{Pd(OAc)_2,\ K_2CO_3,\ CuI} Ar_2NH-C_6H_4-Y \qquad (5.49)$$

Ar$_2$NH≡Ph$_2$NH, Carbasole; Y = H, *p*-Cl, *p*-Me, *p*-Meo

of diethyl hydrogen phosphite with aryl iodides and bromides generated arylphosphonates [Eq. (5.50)] (162). The reaction occurs either with or without the use of a phosphine ligand.

$$(EtO)_2P(O)H + Ar—X \xrightarrow[\substack{H_2O, Bu_4NCl, 60°C, 4 h, \\ 90–99\%}]{Pd(OAc)_2, K_2CO_3, CuI} (EtO)_2P(O)—Ar \qquad (5.50)$$

X = I, Br; "Pd" = Pd(OAc)$_2$(m-Ph$_2$PC$_6$H$_4$SO$_3$H)$_2$, PdCl$_2$, Pd(OAc)$_2$

Other transition metals have also been used. For example, Trost (163) reported that heating a 1:1 mixture of 1-octene and 1-octyne in DMF–water (3:1) at 100°C with a ruthenium complex for 2 h generated a 1:1 mixture of two products corresponding to the addition of the alkene to the acetylene [Eq. (5.51)]. The presence of a normally reactive enolate does not interfere with the reaction [Eq. (5.52)].

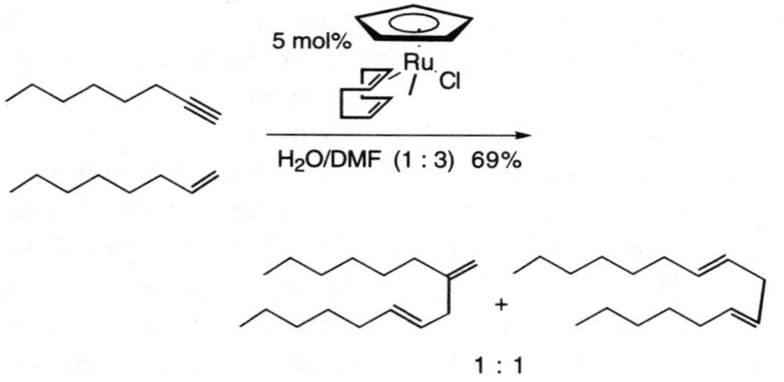

(5.51)

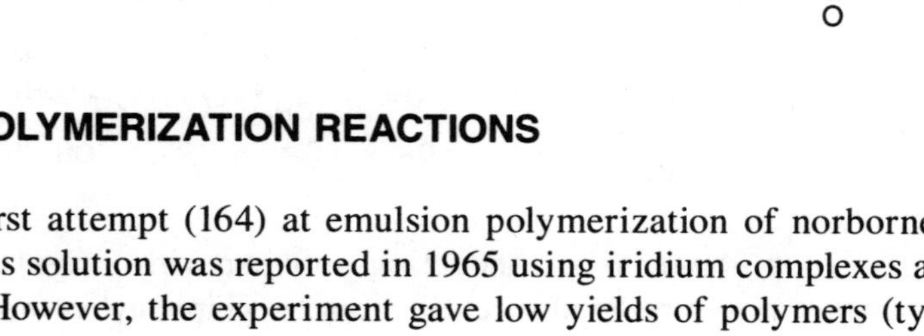

(5.52)

5.6 POLYMERIZATION REACTIONS

The first attempt (164) at emulsion polymerization of norbornenes in aqueous solution was reported in 1965 using iridium complexes as catalysts. However, the experiment gave low yields of polymers (typically

<10%). Recently, Novak and Grubbs reported that 7-oxanorbornene derivatives rapidly polymerized in aqueous solution under air atmosphere using some selected group VIII coordination complexes as catalysts, providing a quantitative yield of a ring-opening metathesis polymerization (ROMP) product [Eq. (5.53)] (165). A similar ring-opening metathesis polymerization reaction occurred with norbornene (166). The ruthenium catalyst used in this polymerization can be prepared by the Bernhard–Ludi procedure (167).

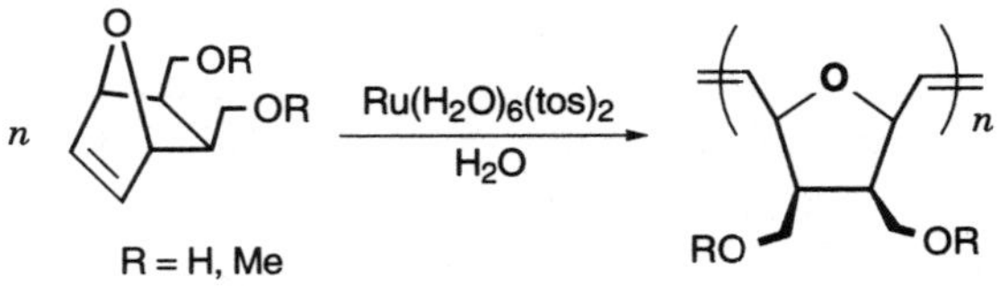

$$(5.53)$$

Compared with the same reaction carried out in organic solvent, the initiation time decreased from 22–24 h to 30–35 min. After the polymerization, the aqueous catalyst solution not only could be reused but also became more active in subsequent polymerizations, and the initiation period dropped to only 10–12 s. Solutions containing these aqueous catalysts have been recycled for 14 successive polymerizations without any detectable loss of activity. Wache recently described the use of Ru(IV) complexes, $[Ru(\eta^3: \eta^3\text{-}C_{10}H_{16})(H_2O)(OAc)]^+BF_4^-$ and $[Ru(\eta^3: \eta^3\text{-}C_{10}H_{16})(H_2O)(OTf)_2]$, for the emulsion ring-opening polymerization of norbornene (168). The polymers generated had very high molecular weights ($M_n \sim 1.5 \times 10^6$ g/mol^{-1}) and high *cis* selectivity (85–90%).

By using the Ru-catalyzed ring-opening polymerization in water, Kiessling synthesized neoglycopolymers for intercellular recognition studies (169):

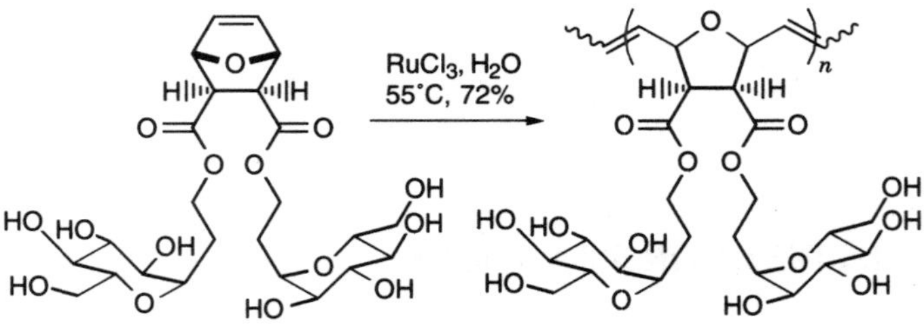

$$(5.54)$$

A hydrocarbon nonmetallic conducting polymer with a rigid rod of benzene rings was synthesized recently from two biphenyl compounds via the Suzuki reaction in water (170). The reaction was catalyzed by a water-soluble palladium catalyst [Eq. (5.55)]. Water-soluble palladium compounds have also been used by Sen for the alternating copolymerization of olefins with carbon monoxide in aqueous media (171).

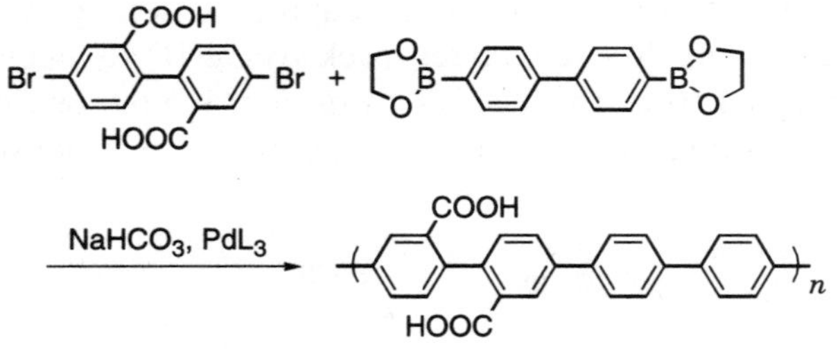

$$L = P(C_6H_5)_2(m\text{-}C_6H_4SO_3Na)$$

(5.55)

By using the palladium-catalyzed coupling between aryl halides with acetylene gas, a variety of poly(arene ethynylene)s was prepared by Li et al. (172) from aryl diiodides. A helical polymer was similarly generated from a diiodobinaphthol derivative:

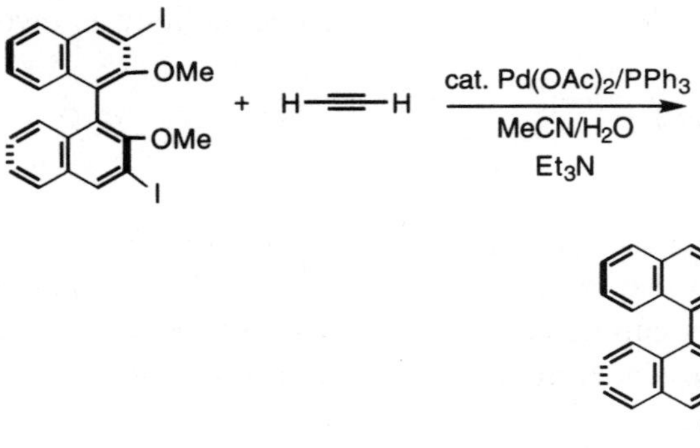

(5.56)

5.7 TRANSITION-METAL-CATALYZED OXIDATION REACTIONS

Among transition-metal-catalyzed oxidations in aqueous media, the most important one is the palladium-catalyzed oxidation of alkenes. The oxidation of ethylene to acetaldehyde by palladium chloride in water has been known since the nineteenth century (173). However, the reaction requires the use of a stoichiometric amount of $PdCl_2$, resulting in Pd(0) deposit. Anderson, in 1934, observed a similar (but much less efficient) reaction with the Zeise salt in water at a higher temperature (174). In the late 1950s, Smidt (175) of Wacker Chemie discovered that by using $CuCl_2$, Pd(0) can be oxidized back to Pd(II) before it deposits out. The $CuCl_2$ itself is reduced to cuprous chloride, which is air-sensitive, and is readily reoxidized back to Cu(II) (Scheme 5.8). The reaction, referred to as the *Wacker oxidation*, is now one of the most important transition-metal-catalyzed reactions in the industry.

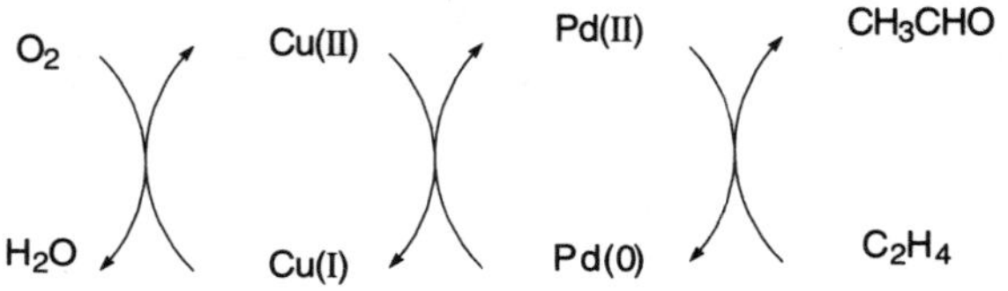

Scheme 5.8

The mechanism for the product formation was originally proposed to involve a *cis* transfer of OH to a metal-bound alkene (176). However, stereochemical studies carried out by Bäckvall (177) and by Stille (178) proved otherwise, demonstrating that a *trans* addition is involved:

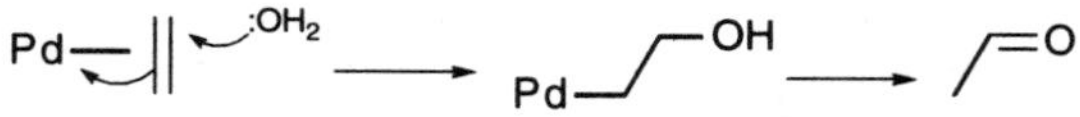

$$(5.57)$$

For internal olefins, the Wacker oxidation is sometimes surprisingly regioselective. By using aqueous dioxane or THF, oxidation of β,γ-unsaturated esters can be achieved selectively to generate γ-ketoesters [Eq. (5.58)] (179). Under appropriate conditions, Wacker oxidation can

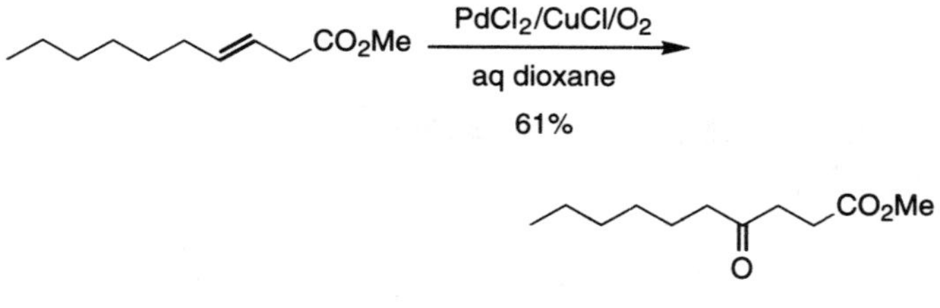

$$(5.58)$$

be used very efficiently in transforming an olefin to a carbonyl compound. Thus, olefins become masked ketones. An example is the application of Wacker oxidation in the synthesis of $(+)$-19-nortestosterone (**17**) carried out by Tsuji (Scheme 5.9) (180).

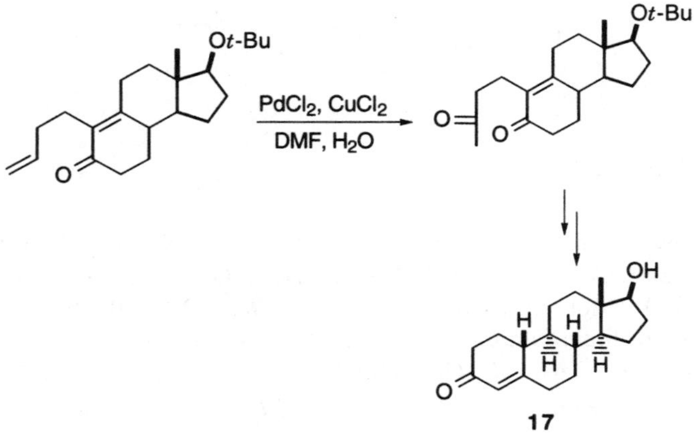

Scheme 5.9

Besides Wacker oxidation, other transition-metal-catalyzed oxidations have also been carried out in aqueous media. For example, methyl groups can be selectively hydroxylated by platinum salts in water (181). In this way, p-toluenesulfonic acid was oxidized to the benzylic alcohol, which was subsequently oxidized into the aldehyde (182):

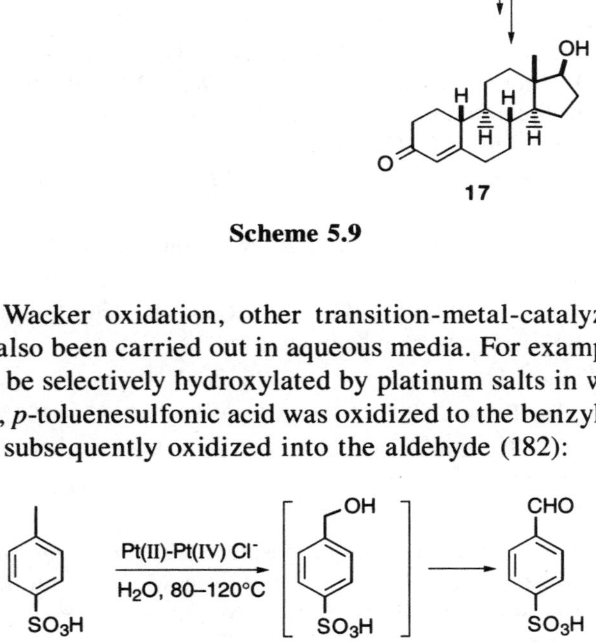

$$(5.59)$$

Oxidation of simple alcohols (cyclohexanol, benzyl alcohol) to the corresponding aldehydes or ketones can be achieved in water using water stable catalysts **18–20** and an oxidant, tert-butylhydroperoxide

$[Fe_2O(\eta^1\text{-}H_2O)(\eta1\text{-}OAc)(TPA)_2]^{3+}$
18 TPA = tris[(2-pyridyl)methyl]amine

$[Fe_2O(\eta^1\text{-}H_2O)(\eta1\text{-}OAc)(BPIA)_2]^{3+}$
19 BPIA = bis[(2-pyridyl)methyl][2-(1-methylimidazoyl)methyl]amine

$[Fe_2O(\eta^1\text{-}H_2O)_2(BIPA)_2]^{4+}$
20 BIPA = bis[2-(1-methylimidazoyl)methyl][(2-pyridyl)methyl]amine

(TBHP) [Eq. (5.60)] (183). The reaction serves as a biomimetic oxidation for methane monooxygenase enzyme (MMO). More examples of oxidations will be discussed in the next chapter.

$$\text{(5.60)}$$

REFERENCES

1. For reviews, see Joo, F.; Toth, Z. *J. Mol. Catal.* **8**, 369 (1980); Sinou, D. *Bull. Soc. Chim. Fr.* 480 (1987); Kuntz, E. G. *Chemtech* **17**, 570 (1987); Kalck, P.; Monteil, F. *Adv. Organomet. Chem.* **34**, 219 (1992); Hermann, W. A.; Kohlpainter, C. W. *Angew. Chem., Int. Ed. Engl.* **32**, 1524 (1993); Roundhill, D. M. *Adv. Organomet. Chem.* **34**, 155 (1995).

2. Chatt, J.; Leigh, G. J.; Slade, R. M. *J. Chem. Soc. Dalton Trans.* 2021 (1973).

3. Ahrland, S.; Chatt, J.; Davis, N. R.; Williams, A. A. *J. Chem. Soc.* 276 (1958).

4. Kuntz, E. G. Fr. Patent 2,314,910 to Rhone-Poulenc Industries (06-20-1975), modifications have been reported (see Ref. 1).

5. Schindlbauer, H. *Monatsch. Chem.* **96**, 2051 (1965).

6. Gilman, H.; Brown, G. E. *J. Am. Chem. Soc.* **67**, 824 (1945).

7. Luckenbach, R.; Lorenz, K. *Z. Naturforsch. B* **32**, 1038 (1977).

8. Okano, T.; Morimoto, K.; Konishi, H.; Kiji, J. *Nippon Kagaku Kaishi* 486 (1985).

9. Nuzzo, R. G.; Feitler, D.; Whitesides, G. M. *J. Am. Chem. Soc.* **101**, 3683 (1979); Nuzzo, R. G.; Haynie, S. L.; Wilson, M. E.; Whitesides, G. M. *J. Org. Chem.* **46**, 2861 (1981); Wlison, M. E.; Nuzzo, R. G.; Whitesides, G. M. *J. Am. Chem. Soc.* **100**, 2270 (1978).

10. Fell, B.; Pagadogianakis, *J. Mol. Catal.* **66**, 143 (1991).

11. Dibowski, H.; Schmidtchen, F. P. *Tetrahedron Lett.* **51**, 2325 (1995).

12. Lavenot, L.; Bortoletto, M. H.; Roucoux, A.; Larpent, C.; Patin, H. *J. Organomet. Chem.* **509**, 9 (1996).

13. Ding, H.; Hanson, B. E.; Bartik, T.; Bartik, B. *Organometallics* **13**, 3761 (1994).

14. Herrmann, W. A.; Albanese, G. P.; Manetsberger, R. B.; Lappe, P.; Bahrmann, H. *Angew. Chem., Int. Ed. Engl.* **34**, 811 (1995).

15. Herrmann, W. A.; Kohlpainter, C. W.; Bahrmann, H.; Konkol, W. *J. Mol. Catal.* **73**, 191 (1992).

16. Herrmann, W. A.; Kohlpainter, C. W.; Manetsberger, R. B.; Bahrmann, H. (Hoechst AG), DE-B 4220267A, 1992.

17. Alario, F.; Amrani, Y.; Colleuille, Y.; Dang, T. P.; Jenck, J.; Morel, D.; Sinou, D. *J. Chem. Soc., Chem. Commun.* 202 (1986).

18. Barton, M.; Atwood, J. D. *J. Coord. Chem.* **24**, 43 (1991).

19. Smith, R. T.; Baird, M. C. *Inorg. Chim. Acta* **62**, 135 (1982).

20. Ganguly, S.; Mague, J. T.; Roundhill, D. M. *Inorg. Chem.* **31**, 3500 (1992).

21. Parshall, G. W. *Homogeneous Catalysis*, Wiley, New York, 1980.

22. Mann, F. G.; Millar, I. T. *J. Chem. Soc.* 4453 (1952).

23. Podlahova, J.; Podlaha, J. *Collect. Czech. Chem. Commun.* **45**, 2049 (1980).

24. Okano, T.; Iwahara, M.; Konishi, H.; Kiji, J. *J. Organomat. Chem.* **346**, 267 (1988).

25. Hoye, P. A. T.; Pringle, P. G.; Smith, M. B.; Worboys, K. *J. Chem. Soc. Dalton Trans.* 269 (1993).

26. Komiya, S.; Awata, S.; Ishimatsu, S. *Inorg. Chim. Acta* **212**, 201 (1994).

27. Avey, A.; Schut, D. M.; Weakley, T. J. R.; Tyler, D. R. *Inorg. Chem.* **32**, 233 (1993).

28. Renaud, E.; Russell, R. B.; Fortier, S.; Brown, S. J.; Baird, M. C. *J. Organomet. Chem.* **419**, 403 (1991).

29. Schull, T. L.; Feltinger, J. C.; Knight, D. A. *J. Chem. Soc., Chem. Commun.* 1487 (1995).

30. Daigle, D. J.; Pepperman, A. B. Jr.; Vail, S. L. *J. Heterocyclic Chem.* **11**, 407 (1974).

31. Amrani, Y.; Sinou, D. *J. Mol. Catal.* **24**, 231 (1984).

32. Alcock, N. W.; Brown, J. M.; Jeffery, J. C. *J. Chem. Soc., Chem. Commun.* 829 (1974).

33. Benhamza, R.; Amrani, Y.; Sinou, D. *J. Organomet. Chem.* **288**, C37 (1985).

34. Toth, I.; Hanson, B. E. *Organometallics* **12**, 1507 (1993).

35. Amrani, Y.; Lecomte, L.; Sinou, D.; Bakos, F.; Toth, I.; Heil, B. *Organometallics* **8**, 542 (1989).

36. Amrani, Y. Ph.D. thesis, University of Lyon, 1986.

37. Amrani, Y.; Sinou, D. *J. Mol. Catal.* **36**, 319 (1986).

38. Ding, H.; Hanson, B. E.; Bakos, J. *Angew. Chem., Int. Ed. Engl.* **34**, 1645 (1995).

39. Wilson, M. E.; Whitesides, G. M. *J. Am. Chem. Soc.* **100**, 306 (1978).

40. Nagel, U.; Kinzel, E. *Chem. Ber.* **119**, 1731 (1986).

41. Hayashi, T.; Yamamoto, K.; Kumada, M. *Tetrahedron Lett.* 4405 (1974).

42. Malmstrom, T.; Andersson, C. *J. Chem. Soc. Chem. Commun.* 1135 (1996).

43. Ahrland, S.; Chatt, J.; Davis, N. R.; Williams, A. A. *Nature* **179**, 118 (1957); *J. Chem. Soc.* 276 (1958).

44. Herrmann, W. A.; Thiel, W. R.; Kuchler, J. G. *Chem. Ber.* **123**, 1953 (1990).

45. Anderson, S.; Constable, E. C.; Seddon, K. R.; Turp, J. E.; Baggott, J. E.; Pilling, M. J. *J. Chem. Soc., Dalton Trans.* 2247 (1985).

46. Sprintschnik, G.; Sprintschnik, H. W.; Kirsch, P. P.; Whitten, D. G. *J. Am. Chem. Soc.* **99**, 4947 (1977).

47. Baker, P. K.; Jenkins, A. E.; Lavery, A. J.; Muldoon, D. J.; Shawcross, A. *J. Chem. Soc. Dalton Trans.* 1525 (1995).

48. Ford, P. C.; Rokicki, A. *Adv. Organomet. Chem.* **28**, 139 (1988).

49. For representative monographs, see Augustine, R. L. *Catalytic Hydrogenation*, Marcel Dekker, New York, 1965; Freifelder, M. *Practical Hydrogenation*, Wiley, New York, 1971; Freifelder, M. *Catalytic Hydrogenation in Organic Synthesis*, Wiley-Interscience, New York, 1978; Rylander, P. N. *Catalytic Hydrogenation in Organic Synthesis*, Academic Press, New York, 1979; Rylander, P. N. *Hydrogenation Methods*, Academic Press, Orlando, 1985.

50. Young, J. F.; Osbourne, J. A.; Jardine, F. H.; Wilkinson, G. *J. Chem. Soc., Chem. Commun.* 131 (1965); Bennett, M. A.; Longstaff, P. A. *Chem. Ind.* (London), 849 (1965); Coffey, R. S.; Smith, J. B. U.K. Patent, 1,121, 642 (1965).

51. Spencer, M. S.; Dowden, D. A. U.S. Patent 3,009,969 (1961); Kwiatek, J.; Madok, I. L.; Syeler, J. K. *J. Am. Chem. Soc.* **84**, 304 (1962).

52. Chatt, J.; Leigh, G. J.; Slade, R. M. *J. Chem. Soc. Dalton Trans.* 2021 (1973).

53. Joo, F.; Beck, M. T. T. *React. Kinet. Catal. Lett.* **2**, 257 (1975).

54. Dror, Y.; Manassen, J. *J. Mol. Catal.* **2**, 219 (1977).

55. Borowski, A. F.; Cole-Hamilton, D. J.; Wilkinson, G.; *Nonv. J. Chim.* **2**, 137, (1978).

56. Larpent, C.; Dabard, R.; Patin, H. *Tetrahedron Lett.* **28**, 2507 (1987).

57. Larpent, C.; Patin, H. *J. Organomet. Chem.* **335**, C13 (1987).

58. Grosselin, J. M.; Mercier, C.; Allmang, G.; Grass, F. *Organometallics* **10**, 2126 (1991).

59. Smith, R. T.; Ungar, R. K.; Sanderson, L. J.; Baird, M. C. *Organometallics* **2**, 1138 (1983).

60. Okano, T.; Kaji, M.; Isotani, S.; Kiji, J. *Tetrahedron Lett.* **33**, 5547 (1992).

61. Tour, J. M.; Pendalwar, S. L. *Tetrahedron Lett.* **31**, 4719 (1990); Tour, J. M.; Cooper, J. P.; Pendalwar, S. L. *J. Org. Chem.* **55**, 3452 (1990).

62. Joo, F.; Benyei, A. *J. Organomet. Chem.* **363**, C19 (1989); Benyei, A.; Joo, F. *J. Mol. Catal.* **58**, 151 (1990).

63. Joo, F.; Toth, Z.; Beck, M. T. *Inorg. Chim. Acta* **25**, L61 (1977).

64. Grosselin, J. M.; Mercier, C. *J. Mol. Catal.* **63**, L25 (1990).

65. Fache, E.; Senocq, F.; Santini, C.; Basset, J. M. *J. Chem. Soc., Chem. Commun.* 1776 (1990).

66. Toth, I.; Hanson, B. E.; Davis, M. E. *Tetrahedron Asymmetry* **1**, 913 (1990).

67. Bakos, J.; Orosz, A.; Heil, B.; Laghmari, M.; Lhoste, P.; Sinou, D. *J. Chem. Soc., Chem. Commun.* 1684 (1991); Lensink, C.; de Vries, J. G. *Tetrahedron Asymmetry* **3**, 235 (1992).

68. Lecomte, L.; Sinou, D.; Bakos, J.; Toth, I.; Heil, B. *J. Organomet. Chem.* **370**, 277 (1989).

69. Halpern, J. in *Asymmetric Synthesis,* Vol. 5, Morrison, J. D., ed. Academic Press, Orlando, 1985.

70. Laghmari, M.; Sinou, D. *J. Mol. Catal.* **66**, L15 (1991).

71. McQuillin, F. J.; Parker, D. G.; Stephenson, G. R. *Transition Metal Organometallics for Organic Synthesis,* Cambridge Press, 1991; Crabtree, R. H. *The Organometallic Chemistry of Transition Metals,* 2nd ed., Wiley, New York, 1994.

72. McGrath, D. V.; Grubbs, R. H. *Organometallics* **13**, 224 (1994).

73. Li, C. J.; Wang, D.; Chen, D. L. *J. Am. Chem. Soc.* **117**, 12867 (1995).

74. For a recent review, see Khan, M. M. T. *Platinum Metals Rev.* **35**(2), 70 (1991).

75. Cornils, B. *Hydroformylation, Oxo Synthesis, Roelen Reaction: New Synthesis with Carbon Monoxide*, Springer-Verlag, Berlin, 1980, pp. 1–224.

76. Roelen, D., D.E. 84584, 1938; Ruhrchemie.

77. Ciardelli, F.; Braca, G.; Carlini, C.; Sbrana, G.; Valentini, G. *J. Mol. Catal.* **14**, 1 (1982).

78. Bemi, L.; Clark, H. C.; Davies, J. A.; Fyfe, C. A.; Wasylishen, R. E. *J. Am. Chem. Soc.* **104**, 438 (1982).

79. Borowski, A. F.; Cole-Hamilton, D. J.; Wilkinson, G. *Nouv. J. Chim.* **2**, 137 (1978).

80. Smith, R. T.; Ungar, R. K.; Baird, M. C. *Trans. Met. Chem.* **7**, 288 (1982).

81. Bahrmann, H.; Bach, H. *Phosphorus Sulfur* **30**, 611 (1987).

82. Kuntz, E., U.S. Patent 4,248,802, Rhone-Poulenc Ind., 1981; *Chem. Abstr.* **87**, 101944n (1977).

83. Jenck, J.; Fr. Patent 2,478,078 to Rhone-Poulenc Industries (03-12-1980); Kuntz, E. Fr. Patent 2,349,562 to Rhone-Poulenc Industries (04-29-1976).

84. Herrmann, W. A.; Kohlpainter, C. W.; Manetsberger, R. B.; Bahrmann, H. (Hoechst AG), DE-B 4220,267A, 1992.

85. Kalck, P.; Escaffre, P.; Serein-Spirau, F.; Thorez, A.; Besson, B.; Colleuille, Y.; Perron, R. *New. J. Chem.* **12**, 687 (1988).

86. Smith, R. T.; Ungar, R. K.; Sanderson, L. J.; Baird, M. C. *Organometallics* **2**, 1138 (1983).

87. Russell, M. J. H.; Murrer, B. A. U.S. Patent 4,399,312 to Johnson Matthey Company (08-27-1981); Russell, M. J. H. *Platinum Met. Rev.* **32**, 179 (1988).

88. Fell, B.; Pagadogianakis, G. *J. Mol. Catal.* **66**, 143 (1991).

89. Monflier, E.; Fremy, G.; Castanet, Y.; Mortreux, A. *Angew. Chem., Int. Ed. Engl.* **34**, 2269 (1995).

90. Anderson, J. R.; Campi, E. M.; Jackson, W. R. *Catal. Lett.* **9**, 55 (1991).

91. Arhancet, J. P.; Davis, M. E.; Merola, J. S.; Hanson, B. E. *Nature* **339**, 454 (1989); Haggin, J. *Chem. Eng. News*, **70**(17), 40 (1992); Herrmann, W. A. *Hoechst High Chem. Magazine* No. 13, 14 (1992).

92. Khna, M. M. T.; Halligudi, S. B.; Abdi, S. H. R. *J. Mol. Catal.* **48**, 313 (1988).

93. Jenner, G. *Tetrahedron Lett.* **32**, 505 (1991).

94. Heck, R. F. *Palladium Reagents in Organic Syntheses*, Academic Press, London, 1985.

95. Kiji, J.; Okano, T.; Nishiumi, W.; Konishi, H. *Chem. Lett.* 957 (1988); see also references cited therein.

96. Alper, H.; Abbayes, H. D. *J. Organomet. Chem.* **134**, C11 (1977).

97. Cassar, L.; Foa, M. *J. Organomet. Chem.* **134**, C15 (1977).

98. Alper, H. *J. Organomet. Chem.* **300**, 1 (1986); see also references cited therein.

99. Joo, F.; Alper, H. *Organomettallics* **4**, 1775 (1985).

100. Bumagin, N. A.; Nikitin, K. V.; Beletskaya, I. P. *J. Organomet. Chem.* **358**, 563 (1988).

101. Okano, T.; Hayashi, T.; Kiji, J. *Bull. Chem. Soc. Jpn.* **67**, 2339 (1994).

102. Papadogianakis, G.; Maat, L.; Sheldon, R. A. *J. Chem. Soc., Chem. Commun.* 2659 (1994).

103. Bumagin, N. A.; Nikitin, K. V.; Beletskaya, I. P. *J. Organomet. Chem.* **358**, 563 (1988).

104. Gaushin, V. V.; Alper, H. H. *J. Org. Chem.* **58**, 4798 (1993).

105. Monteil, F.; Kalck, P. *J. Organomet. Chem.* **482**, 45 (1994).

106. Shim, S. C.; Antebi, S.; Alper, H. *J. Org. Chem.* **50**, 147 (1985).

107. Urata, H.; Kosukegawa, O.; Tshii, Y.; Yugari, H.; Fuchikami, T. *Tetra-hedron Lett.* **30**, 4403 (1989).

108. Joh, T.; Doyama, K.; Onitsuka, K.; Shiohara, T.; Takahashi, S. *Organometallics* **10**, 2493 (1991).

109. Joh, T.; Nagata, H.; Takahashi, S. *Chem. Lett.* 1305 (1992).

110. Alper, H.; Arzoumanian, H.; Petrignani, J. F.; Saldana-Maldonado, M. *J. Chem. Soc., Chem. Commun.* 340 (1985).

111. Alper, H. ; Currie, J. K.; des Abbayes, H. *J. Chem. Soc., Chem. Com-mun.* 311 (1978); Galamb, V.; Gopal, M.; Alper, H. *J. Chem. Soc., Chem. Commun.* 1154 (1983).

112. Galamb, V.; Gopal, M.; Alper, H. *Organometallics* **2**, 801 (1983).

113. Zargarian, D.; Alper, H. ibid. **10**, 2914 (1991).

114. Khan, M. M.T.; Halligudi, S. B.; Abdi, S. H. R. *J. Mol. Catal.* **48**, 325 (1988).

115. Lin, M.; Sen, A. *J. Am. Chem. Soc.* 1992, *114*, 7307; *Nature* **368**, 613 (1994).

116. Garves, K. *J. Org. Chem.* **35**, 3273 (1970).

117. Trost, B. M.; Verhoeven, T. R. *J. Am. Chem. Soc.* **102**, 4730 (1980).

118. Safi, M.; Sinou, D. *Tetrahedron Lett.* **32**, 2025 (1991); Genet, J. P.; Blart, E.; Savignac, M. *Synlett.* 715 (1992).

119. Sigismondi, S.; Sinou, D.; Perez, M.; Moreno-Manas, M.; Pleixats, R.; Villarroya, M. *Tetrahedron Lett.* **35**, 7085 (1994).

120. Blart, E.; Genet, J. P.; Safi, M.; Savignac, M.; Sinou, D. *Tetrahedron* **50**, 505 (1994).

121. Genet, J. P.; Blart, E.; Savignac, M.; Lemeune, S.; Paris, J. M. *Tetrahe-dron Lettt.* **34**, 4189 (1993).

122. Ruth, J. L.; Bergstrom, D. E. *J. Am. Chem. Soc.* **98**, 1587 (1976); *J. Org. Chem.* **43**, 2870 (1978).

123. (a) Bigge, C. F.; Kalaritis, P.; Deck, J. R.; Mertes, M. P. *J. Am. Chem. Soc.* **102**, 2033 (1980); (b) Bigge, C. F.; Kalaritis, P.; Mertes, M. P. *Te-trahedron Lett.* 1652 (1979).

124. Langer, P. R.; Waldrop, A. A.; Ward, D. C. *Proc. Natl. Acad. Sci.* (USA) **78**, 6633 (1981).

125. For reviews, see Heck, R. F. *Acc. Chem. Res.* **12**, 146 (1979).

126. (a) Bumagin, N. A.; Andryuchova, N. P.; Beletskaya, I. P. *Izv. Akad. Nauk SSSr,* **6**, 1449 (1988); (b) Bumagin, N. A.; More, P. G.; Beletskaya, I. P. *J. Organomet. Chem.* **371**, 397 (1989); Bumagin, N. A.; Bykov, V. V.; Sukhomlinova, L. I.; Tolstaya, T. P.; Beletskaya, I. P. *J. Organomet. Chem.* **486**, 259 (1995).

127. Genet, J. P.; Blart, E.; Savignac, M. *Synlett.* 715 (1992).

128. Bumagin, N. A.; Sukhomlinova, L. I.; Banchikov, A. N.; Tolstaya, T. P.; Beletskaya, I. P. *Bull. Russ. Acad. Sci.* **41**, 2130 (1992).

129. Jeffery, T. *Tetrahedron Lett.* **35**, 3051 (1994).

130. Hiroshige, M.; Hauske, J. R.; Zhou, P. *Tetrahedron Lett.* **36**, 4567 (1995).

131. Reardon, P.; Metts, S.; Crittendon, C.; Daugherity, P.; Parsons, E. J. *Organometallics* **14**, 3810 (1995).

132. Diminnie, J.; Metts, S.; Parsons, E. J. *Organometallics* **14**, 4023 (1995).

133. For reviews, see Stille, J. F. *Angew. Chem., Int. Ed. Engl.* **25**, 508 (1986).

134. Tueting, D. R.; Echavarren, A. M.; Stille, J. K. *Tetrahedron* **45**, 979 (1989).

135. Zhang, H. C.; Davis, Jr. G. D. *Organometallics* **12**, 1499 (1992).

136. Rai, R.; Aubrecht, K. B.; Collum, D. B. *Tetrahedron Lett.* **36**, 3111 (1995).

137. Bumagin, N. A.; Sukhomlinova, L. I.; Tolstaya, T. P.; Beletskaya, I. P. *Russ. J. Org. Chem.* **30**, 1605 (1994).

138. Miyaura, N.; Ishiyama, T.; Sasaki, H.; Ishikawa, M.; Satoh, M.; Suzuki, A. *J. Am. Chem. Soc.* **111**, 314 (1989); for a recent review, see Miyaura, N.; Suzuki, A. *Chem. Rev.* **95**, 2457 (1995).

139. Colberg, J. C.; Rane, A.; Vaquer, J.; Soderquist, J. A. *J. Am. Chem. Soc.* **115**, 6065 (1993).

140. Armstrong, R. W.; Beau, J. M.; Cheon, S. H.; Christ, W. J.; Fujioka, H.; Ham, W. H.; Hawkins, L. D.; Jin, H.; Kang, S. H.; Kishi, Y.; Martinelli, M. J.; McWhorter, W. W.; Mizuno, M.; Nakata, M.; Stutz, A. E.; Talamas, F. X.; Taniguchi, M.; Tino, J. A.; Ueda, K.; Uenishi, J. I.; White, J. B.; Yonaga, M. *J. Am. Chem. Soc.* **111**, 7525 (1989).

141. Roush, W. R.; Brown, B. B. *J. Org. Chem.* **58**, 2162 (1993).

142. NIcolaou, K. C.; Ramphal, J. Y.; Palazon, J. M.; Spanevello, R. A. *Angew. Chem., Int. Ed. Engl.* **27**, 587 (1989).

143. Evans, D. A.; Ng, H. P.; Rieger, D. L. *J. Am. Chem. Soc.* **115**, 11446 (1993).

144. Miyaura, N.; Yanagi, T.; Suzuki, A. *Synth. Commun.* **11**, 513 (1981).

145. Alo, B. I.; Kandil, A.; Patil, P. A.; Sharp, M. J.; Siddiqui, M. A.; Sbieckus, V. *J. Org. Chem.* **52**, 3932 (1987).

146. Casalnuovo, A. L.; Calabrese, J. C. *J. Am. Chem. Soc.* **112**, 4324 (1990).

147. Genet, J. P.; Linquist, A.; Blart, E.; Mouries, V.; Savignac, M.; Vaultier, M. *Tetrahedron Lett.* **36**, 1443 (1995).

148. Grushin, V. V.; Alper, H. *J. Org. Chem.* **57**, 2188 (1992).

149. Cassar, L. *J. Organomet. Chem.* **93**, 253 (1975); Dieck, H. A.; Heck, F. R. *J. Organomet. Chem.* **93**, 259 (1975).

150. Sonogashira, K.; Tohda, Y.; Hagira, N. *Tetrahedron Lett.* 4467 (1975).

151. Piskunov, A. V.; Moroz, A. A.; Shvartsberg, M. S. *Russ. Chem. Bull.* 755 (1987).

152. Calsalnuovo, A. L.; Calabreze, J. C. *J. Am. Chem. Soc.* **112**, 4324 (1990).

153. Genet, J. P.; Blart, E.; Savignac, M. *Synlett.* 715 (1992).

154. Davydov, D. V.; Beletskaya, I. P. *Russian Chem. Bull.* **44**, 965 (1995); Bumagin, N. A.; Bykov, V. V.; Beletskaya, I. P. *Russ. J. Org. Chem.* **31**, 348 (1995).

155. Davydov, D. V.; Beletskaya, I. P. *Russ. Chem. Bull.* **44**, 965 (1995).

156. Vlassa, M.; Tarta, I. C.; Margineanu, F.; Oprean, I. *Tetrahedron* **52**, 1337 (1996).

157. Amatore, C.; Blart, E.; Genet, J. P.; Jutand, A.; Lemaire-Andoire, S.; Savignec, M. *J. Org. Chem.* **60**, 6829 (1995).

158. Li, C. J.; in *Environmentally Benign Organic Synthesis*, P. T. Anastas and T. C., Williamson, eds., Oxford University Press, in press.

159. Roshchin, A. I.; Bumagin, N. A.; Beletskaya, I. P. *Doklady Chem.* **334**, 47 (1994).

160. Davydov, D. V.; Beletskaya, I. P. *Russ. Chem. Bull.* **44**, 1139 (1995).

161. Davydov, D. V.; Beletskaya, I. P. *Russ. Chem. Bull.* **44**, 1141 (1995).

162. Novikova, Z. S.; Demik, N. N.; Agarkov, A. Y.; Beletskaya, I. P. *Russ. J. Org. Chem.* **31**, 129 (1995).

163. Trost, B. M.; Indolese, A. *J. Am. Chem. Soc.* **115**, 4361 (1993).

164. Rhinehart, R. E.; Smith, H. P. *Polym. Lett.* **3**, 1049 (1965).

165. Novak, B. M.; Grubbs, R. H. *J. Am. Chem. Soc.* **110**, 7542 (1988).

166. Nguyen, S. B. T.; Johnson, L. K.; Grubbs, R. H.; Ziller, J. W. *J. Am. Chem. Soc.* **114**, 3974 (1992); Nguyen, S. T.; Grubbs, R. H. *J. Am. Chem. Soc.* **115**, 9858 (1993); Fraser, C.; Grubbs, R. H. *Macromolecules* **28**, 7248 (1995).

167. Bernhard, P.; Lehmann, H.; Ludi, A. *J. Chem. Soc., Chem. Commun.* 1216 (1981); Bernhard, P.; Biner, M.; Ludi, A. *Polyhedron* **9**, 1095 (1990).

168. Wache, S. *J. Organomet. Chem.* **494**, 235 (1995).

169. Mortell, K. H.; Gingras, M.; Kiessling, L. L. *J. Am. Chem. Soc.* **116**, 12053 (1994); Mortell, K. H.; Weatherman, R. V.; Kiessling, L. L. *J. Am. Chem. Soc.* **118**, 2297 (1996).

170. Wallow, T. I.; Novak, B. M. *J. Am. Chem. Soc.* **113**, 7411 (1991).

171. Jiang, Z.; Sen, A. *Macromolecules* **27**, 7215 (1994).

172. Li, C. J.; Wang, D.; Slaven, IV, W. T.; *Tetrahedron Lett.* **37**, 4459 (1996).

173. Phillips, I. C. *J. Am. Chem. Soc.* **16**, 255 (1894).

174. Anderson, J. S. *J. Chem. Soc.* 971 (1934).

175. Smidt, J.; Hafner, W.; Jira, R.; Sedlmeier, J.; Sieber, R.; Ruttinger, R.; Kojer, H. *Angew. Chem.* **71**, 176 (1959); **74**, 93 (1962).

176. Henry, P. M. *Adv. Organomet. Chem.* **13**, 363 (1975).

177. Bäckvall, J. E.; Åkermark, B., Lijunggren, S. O. *J. Am. Chem. Soc.* **101**, 2411 (1979).

178. Stille, J. K.; Divakaruni, R. *J. Am. Chem. Soc.* **100**, 1303 (1978).

179. Nagashima, H.; Sakai, K.; Tsuji, J. *Chem. Lett.* 859 (1982).

180. Tsuji, J.; Shimizu, I.; Suzuki, H.; Naito, Y. *J. Am. Chem. Soc.* **101**, 5070 (1979).

181. Labinger, J. A.; Herring, A. M.; Bercaw, J. E. *J. Am. Chem. Soc.* **112**, 5628 (1990).

182. Luinstra, G. A.; Labinger, J. A.; Bercaw, J. E. *J. Am. Chem. Soc.* **115**, 3004 (1993).

183. Rabin, A.; Chen, S.; Wang, J.; Buchanan, R. M.; Seris, J. L.; Fish, R. H. *J. Am. Chem. Soc.* **117**, 12356 (1995).

CHAPTER 6

OXIDATIONS AND REDUCTIONS

Men of practical knowledge find their gratification among rivers.
—Confucius (551–479 B.C.)*

The use of aqueous media in most oxidations and reductions is in most cases due to the convenience of readily available aqueous reagent solution. However, the use of water as solvent has a special effect in some reactions.

6.1 EPOXIDATIONS

In the natural environment of water and air, biological oxygenations of organic compounds are catalyzed by cytochrome P-450. One of the most important oxygenations by P-450 is the epoxidation of alkenes. Recently, extensive studies have been carried out in synthesizing water-soluble metalloporphyrins as epoxidation and other oxygenation catalysts in aqueous media to mimic the properties of P-450. With the use of these water-soluble metalloporphyrins, epoxidations of alkenes can be accomplished with a variety of oxidizing reagents, such as PhIO, NaClO, O_2, H_2O_2, ROOH, and $KHSO_5$. This area has been reviewed in detail (1). For example, carbamazepine reacted with $KHSO_5$ catalyzed

---

The Analects of Confucius, L. Giles, ed. and tr., The Commercial Press, Shanghai, 1933.

by water-soluble iron and manganese porphyrins (**1** and **2**) in water to give the corresponding expoxide (2):

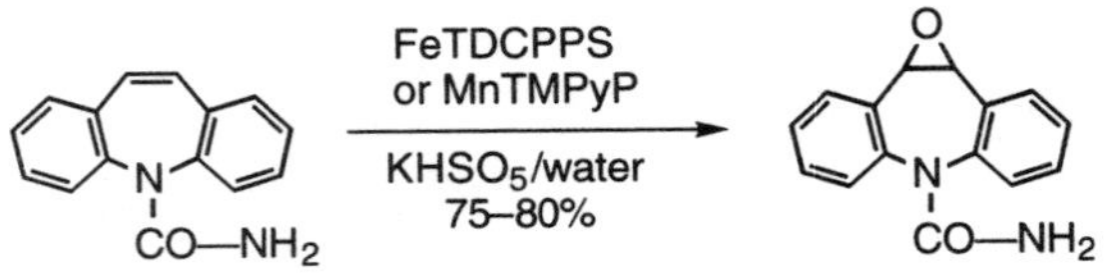

$$(6.1)$$

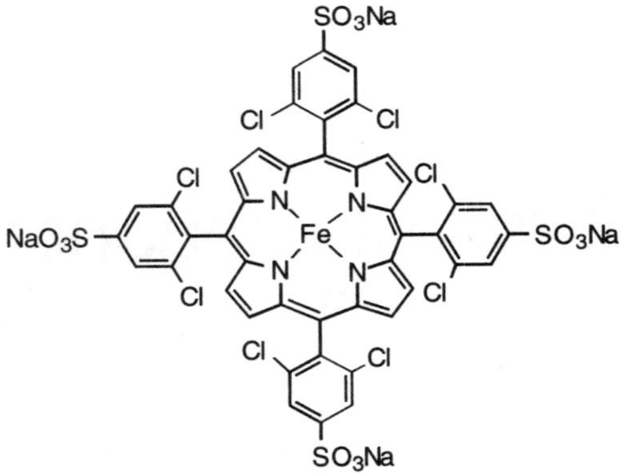

1 FeTDCPPS

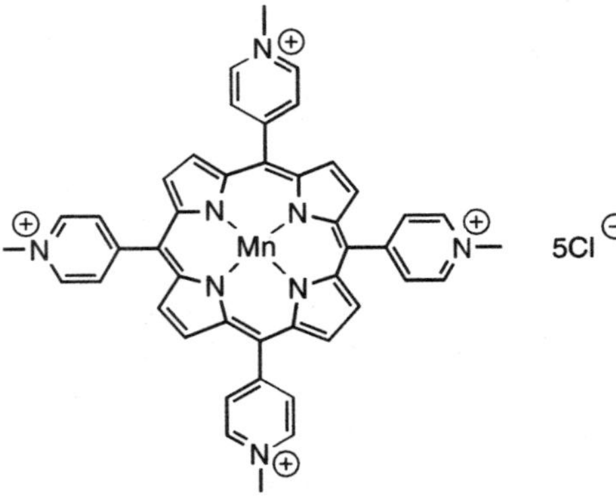

2 MnTMPyP

Alkenes can also be epoxidized directly with a variety of organic peroxy acids or related reagents such as peroxy carboximidic acid, $RC(NH)OOH$, which is readily available through *in situ* reaction of a nitrile with hydrogen peroxide. Thus, reactions of alkenes with *m*-chloroperoxybenzoic acid in water at room temperature give the expoxides in high yields (3). With the use of monoperoxyphthalic acid together with cetyltrimethylammonium hydroxide (CTAOH) as base to control the pH of the aqueous medium, highly regioselective expoxidation of allyl alcohols in the presence of other $C=C$ bonds is possible (4):

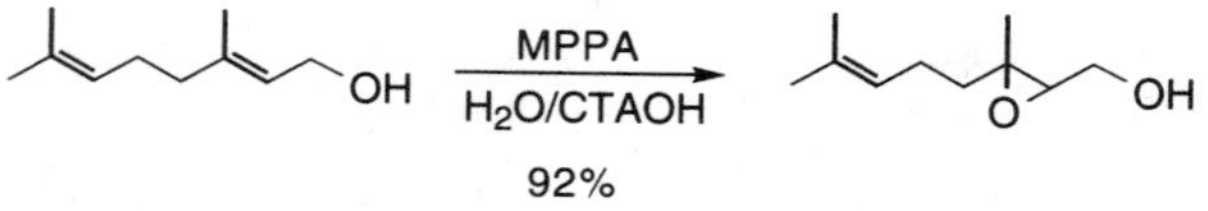

$$\text{(6.2)}$$

Epoxidation of compounds in which a double bond is conjugated to electron-withdrawing groups occurs only very slowly or not at all with peroxy acids or alkyl peroxides. On the other hand, epoxidation with hydrogen peroxide under basic biphase conditions, known as the *Weitz–Scheffer epoxidation* [Eq. (6.3)] (5), is an efficient method for the con-

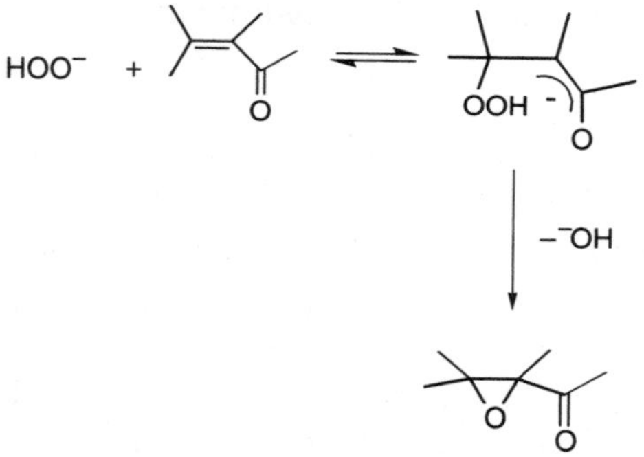

$$\text{(6.3)}$$

version into epoxides. This reaction has been applied to many α,β-unsaturated aldehydes, ketones, nitriles, esters, sulfones, and other compounds. The reaction is first-order in both unsaturated ketone and $^{-}O_2H$ through a Michael-type addition of the hydrogen peroxide anion to the conjugated system followed by ring closure of the intermediate enolate with expulsion of ^{-}OH. The epoxidation of electron-deficient olefins can also be performed with hydrogen peroxide in the presence of sodium tungstate as catalyst (6).

6.2 DIHYDROXYLATIONS

6.2.1 *syn* Dihydroxylation

When the reaction was first discovered, the *syn* dihydroxylation of alkenes was carried out by using a stoichiometric amount of osmium tetroxide in dry organic solvent (7). Hoffman observed that alkenes could react with chlorate salts as the primary oxidants together with a catalytic quantity of osmium tetraoxide, yielding *syn*-vicinal diols [Eq. (6.4)]. This catalytic reaction is usually carried out in water–tetrahydrofuran solvent mixture, and silver or barium chlorate generally give better yields (8).

$$HO_2C\diagdown\diagup CO_2H \xrightarrow[\text{THF–H}_2\text{O, 98\%}]{\text{OsO}_4,\ \text{NaClO}_3} \underset{HO\quad\ OH}{HO_2C\diagup\diagdown CO_2H} \tag{6.4}$$

The *syn* hydroxylation of alkenes is also effected by a catalytic amount of osmium tetroxide in the presence of hydrogen peroxide. Originally developed by Milas, the reaction can be performed with aqueous hydrogen peroxide in solvents such as acetone or diethyl ether (9). Allyl alcohol is quantitatively hydroxylated in water (10):

$$\diagup\!\!\diagup\!\!\diagdown OH \xrightarrow[\text{H}_2\text{O, 100\%}]{\text{OsO}_4,\ \text{H}_2\text{O}_2} HO\diagdown\!\!\diagup\!\!\diagdown OH \tag{6.5}$$

Catalytic dihydroxylations using the osmium tetroxide–*tert*-butyl hydroperoxide system are due largely to Sharpless and co-workers. The aqueous 70% *tert*-butyl hydroperoxide is commercially available and ideal for direct use in this dihydroxylation process (11).

A very effective way of carrying out *syn* dihydroxylation of alkenes is by using an osmium tetroxide–tertiary amine *N*-oxide system. This dihydroxylation is usually carried out in aqueous acetone in either one- or two-phase systems, but other solvents may be required to overcome problems of substrate solubility (12).

Potassium permanganate, usually in alkaline conditions, using aqueous or aqueous–organic solvents, is a widely used oxidant for effecting *syn*-vicinal hydroxylation of alkenes [Eq. (6.6)]. However, overoxidation or alternative oxidation pathways may pose a problem, and the conditions must be carefully controlled (13).

$$\text{norbornene} \xrightarrow[\text{-10°C/3-5 min, 40\%}]{\substack{\text{KMnO}_4\\ \text{H}_2\text{O}/t\text{-BuOH/NaOH}}} \text{diol} \tag{6.6}$$

A photoinduced dihydroxylation of methacryamide by chromium(VI) reagent in aqueous solution was recently reported and may

have potential synthetic applications in the *syn* dihydroxylation of electron-deficient olefins (14). Recently, Minato et al. demonstrated that
$K_3Fe(CN)_6$ in the presence of K_2CO_3 in aqueous *tert*-butyl alcohol provides a powerful system for the osmium-catalyzed dihydroxylation of
olefins (15). This combination overcomes the disadvantages of overoxidation and low reactivity on hindered olefins related to previous
processes:

$$\text{(structure)} \xrightarrow[\substack{K_2CO_3/aq\ t\text{-BuOH} \\ 88\%}]{\text{cat. }OsO_4/K_3Fe(CN)_6} \text{(product)} \tag{6.7}$$

6.2.2 *anti* Dihydroxylation

Hydrogen peroxide in the presence of some oxides, notably tungstic oxide (WO_3) or selenium dioxide (SeO_2), reacts with alkenes to give *anti-*
dihydroxlyation products [Eq. (6.8)] (16). The WO_3 process is best
performed at elevated temperatures (50–70°C) and in aqueous solution.
Aqueous/organic mixed solvents have also been used successfully to
increase the solubility of the alkene in the reaction medium.

$$\underset{R_3 \quad R_4}{\overset{R_1 \quad R_2}{\diagup\!\!\diagdown}} \xrightarrow[H_2O_2]{WO_3 \text{ or } SeO_2} \underset{R_3 \quad OH}{\overset{HO \quad R_2 \quad R_4}{\diagup}} \tag{6.8}$$

It has been proposed that peroxytungstic acid (H_2WO_5) is involved in
the dihydroxylation of alkenes by the hydrogen peroxide–tungstic oxide reagent.

6.2.3 Asymmetric Dihydroxylation

Initially, asymmetric dihydroxylation (17) by osmium tetroxide was
carried out stoichiometrically with chiral diamine ligands (18). Early
attempts to effect dihydroxylation catalytically resulted in low enantiomeric excesses of the products (lower than the stoichiometric methods) (19). The lower optical yield was attributed to the presence of a
second catalytic cycle that exhibits only low or no enantioselectivity.
Sharpless discovered that under aqueous–organic (two-phase) conditions with $K_3Fe(CN)_6$ as the stoichiometric reoxidant and using deriva-

tives of cinchona alkaloids as the chiral ligands, the undesired catalytic pathway can be eliminated, thus resulting in high enantiomeric-excess levels for the desired dihydroxy products (20). Under such conditions, there is no oxidant other than OsO_4 in the organic phase. The addition of 1 equiv of $MeSO_2NH_2$ to the reaction mixture significantly increases the rate of the reaction (21). The use of two independent cinchona alkaloid units attached to a heterocyclic spacer led to further increases of the enantioselectivity and scope of the reaction (22). A ready-made mixture containing all reagents has been commercialized by the name "AD mix."

The stereochemical outcome of the dihydroxylation has been rationalized by a model (Fig. 6.1) (23). An olefin positioned according to

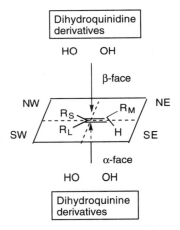

Figure 6.1   Rationale for predicting the enantiofacial selectivity in Sharpless dihydroxylation.

this model will be attacked either from the top face with DHQD (**3**) ligands or from the bottom face with DHQ (**4**) ligands.

3 DHQD   **4** DHQ

For 1,1-disubstituted olefins, the prediction of the stereochemical outcome is based on a modified model (Fig. 6.2) (24).

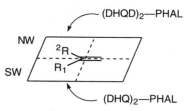

Figure 6.2 Rationale for enantioselectivity of 1,1-disubstituted olefins.

By using the Sharpless dihydroxylation, a variety of compounds have been transformed to diols with high enantiomeric-excess levels. The asymmetric dihydroxylation has a wide range of synthetic applications. As an illustration, the dihydroxylation was used as the key step in the synthesis of squalestatin-1 (**5**) (25):

$$OsO_4, K_3Fe(CN)_6 \quad K_2CO_3, MeSO_2NH_2 \quad t\text{-BuOH–H}_2O \quad DHQD\text{–CLB} \quad RT, 3.5 \text{ days}, 65\%$$

(6.9)

5 squalestatin-1

6.2.4 Aminohydroxylation

A related reaction to asymmetric dihydroxylation is the asymmetric aminohydroxylation of olefins, forming *vic*-aminoalcohols. The β-hydroxyamino group is found in many biologically important molecules, such as the β-aminoacid **6** (the side chain of taxol). In the mid 1970s, Sharpless (26) reported that the trihydrate of N-chloro-p-toluenesulfonamide sodium salt (Chloramine-T) reacts with olefins in the presence of a catalytic amount of osmium tetroxide to produce vicinal hydroxyl p-toluenesulfonamides [Eq. (6.10)]. Aminohydroxylation was also promoted by palladium (27).

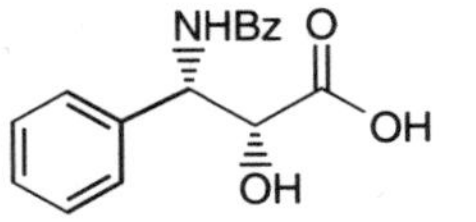

$$(6.10)$$

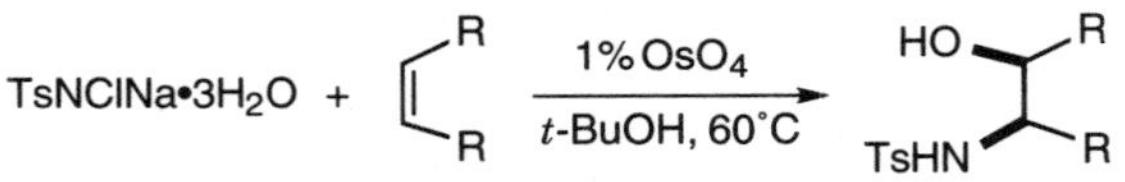

Subsequently, stoichiometric asymmetric aminohydroxylation was reported (28). Recently, Sharpless (29) found that by the combination of chloramine T/OsO$_4$ catalyst with phthalazine ligands used in the asymmetric dihydroxylation reaction, catalytic asymmetric aminohydroxylation of olefins was realized in aqueous acetonitrile or tert-butanol [Eq. (6.11)]. The use of aqueous *tert*-butanol is advantageous

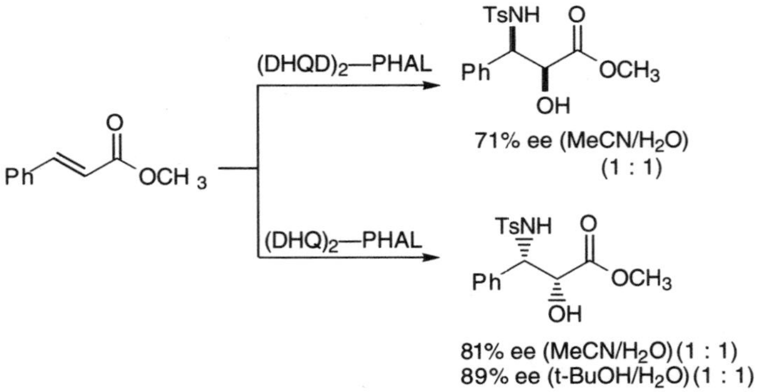

$$(6.11)$$

when the reaction product is not soluble. In this case, essentially pure products can be isolated by a simple filtration, and the toluenesulfonamide byproduct remains in the mother liquor. A variety of olefins were aminohydroxylated in this way (Table 6.1). The reaction is not only performed in aqueous medium but also not sensitive to oxygen. Electron-deficient olefins such as fumarate reacted similarly with high enantiomeric-excess values.

TABLE 6.1 Catalytic Asymmetric Aminohydroxylation

Substrate	Product	ee% (DHD)	ee% (DHQD)
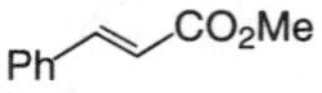 Ph~CO$_2$Me	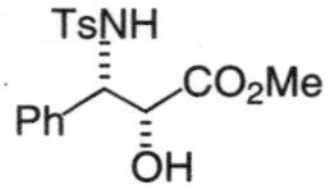 TsNH / Ph~CO$_2$Me / OH	81 (MeCN/H$_2$O) 82 (t-BuOH/H$_2$O)	71
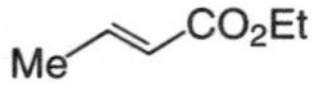 Me~CO$_2$Et	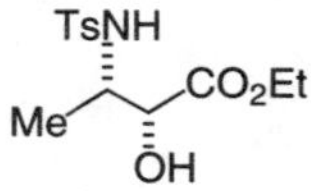 TsNH / Me~CO$_2$Et / OH	74	60
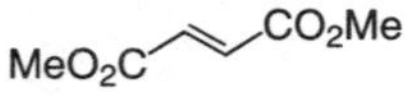 MeO$_2$C~CO$_2$Me	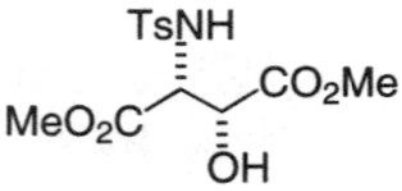 TsNH / MeO$_2$C~CO$_2$Me / OH	77	53
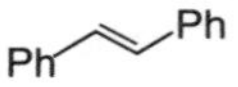 Ph~Ph	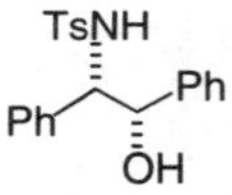 TsNH / Ph~Ph / OH	62 (MeCN/H$_2$O) 78 (t-BuOH/H$_2$O)	50
 Ph~Ph	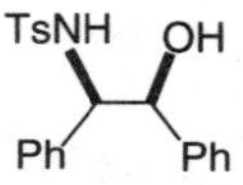 TsNH OH / Ph~Ph	33 (MeCN/H$_2$O) 50 (t-BuOH/H$_2$O)	48
 (cyclohexene)	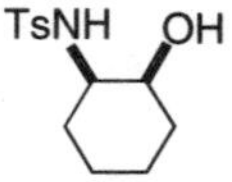 TsNH OH (cyclohexane ring)	45	36

6.3 OXIDATION–DEHYDROGENATION

Because of insolubility of the inorganic reagent in many organic solvents, oxidation by a variety of inorganic oxidizing reagents, such as
potassium permanganate, chromic acid, sodium or potassium dichro-
mate, sodium periodate, and cerium ammonium nitrate (CAN), is often carried out in aqueous medium or in an aqueous–organic biphase
system. For example, CAN has been used to oxidize substituted toluene
to aryl aldehydes. Selective oxidation at one methyl group can be
achieved [Eq. (6.12)] (30). The reaction is usually carried out in aque-

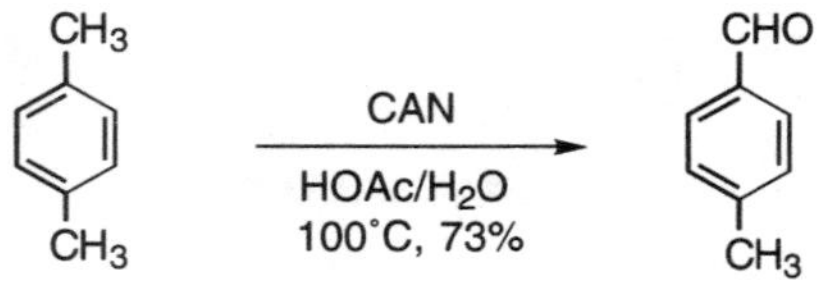

$$
\text{CH}_3\text{-C}_6\text{H}_4\text{-CH}_3 \quad \xrightarrow[\text{HOAc/H}_2\text{O} \atop 100°\text{C, 73\%}]{\text{CAN}} \quad \text{CHO-C}_6\text{H}_4\text{-CH}_3
$$

(6.12)

ous acetic acid. Catalytic Ce(IV) reagent can be used when an additional oxidant, such as sodium bromate, is present. Hydroquinones, cat-
echols, and their derivatives are oxidized to quinones by this system in
aqueous acetonitrile [e.g., Eq. (6.13)] (31). While unactivated primary

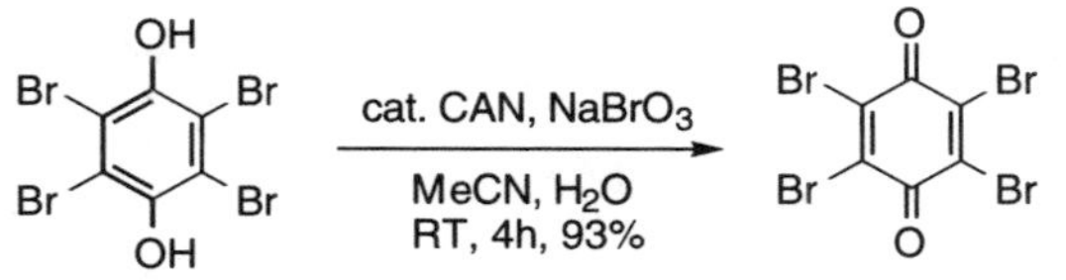

$$(6.13)$$

alcohols are not reactive under the Ce(IV) conditions, a variety of secondary and benzylic alcohols are oxidized by CAN to carbonyl compounds. Tertiary alcohols are oxidatively cleaved to give a variety of
useful products [Eq. (6.14)] (32). The subject has been reviewed in detail by Molander (33).

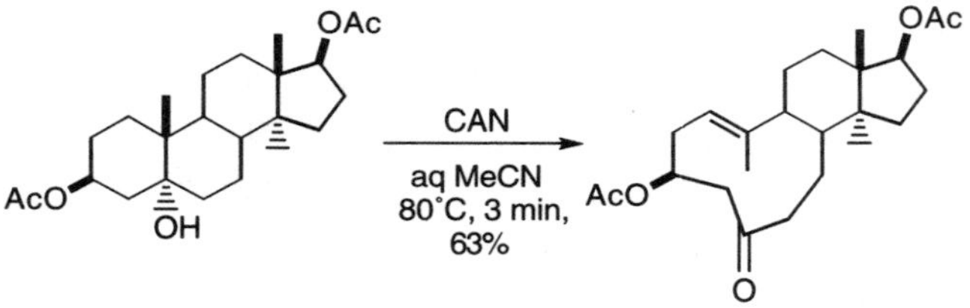

$$(6.14)$$

Oxone is a stable water-soluble oxidant with the approximate com-
position $K_2SO_4 \cdot 2KHSO_5 \cdot KHSO_4$. The reagent has been widely used
for the oxidation of various organic compounds (34). Alkenes, aldehydes, ketones, sulfoxides, sulfides, thiols, disulfides, and phenols are
oxidized by the reagent in water at room temperature (35). The reactions are more rapid in phosphate buffer (pH 6–7) than in pure water.

Several selenium-based oxidizing reagents are routinely used in
aqueous conditions. For example, a mixture of 1,4-dioxane and water is
often used as the solvent for the conversion of aldehydes and ketones by
H_2SeO_3 to α-dicarbonyl compounds in one step [Eq. (6.15)] (36). Dehydrogenation of carbonyl compounds with selenium dioxide generates
the α,β-unsaturated carbonyl compounds in aqueous acetic acid (37).

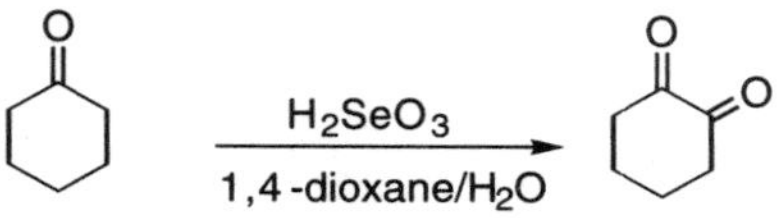

$$(6.15)$$

Dehydrogenation of some cyclic tertiary amines by oxidation with mercuric acetate in aqueous acetic acid leads to the introduction of a carbon–carbon double bond in the α,β-position with respect to the nitrogen atom [Eq. (6.16)] (38). Aqueous bromine has been used as an oxidizing system for many organic transformations (39).

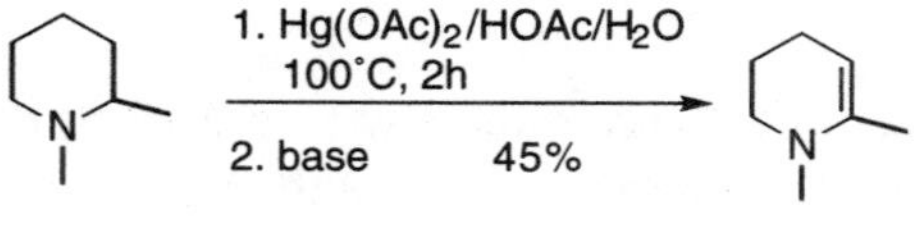

$$
\begin{array}{c}
\text{1. Hg(OAc)}_2\text{/HOAc/H}_2\text{O} \\
\xrightarrow{\hspace{1cm} 100°\text{C, 2h} \hspace{1cm}} \\
\text{2. base} \qquad 45\%
\end{array}
\tag{6.16}
$$

Photochemical oxidative dimerization of capsaicin in aqueous ethanol generated 60% the dimer within 20 min of irradiation [Eq. (6.17)] (40). However, comparison was not available about the reaction in organic solvents.

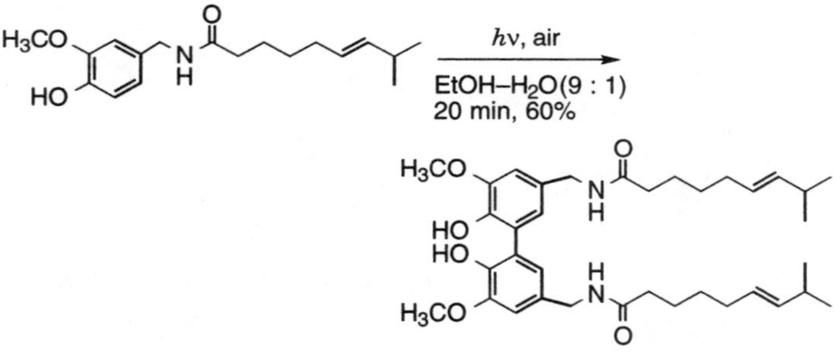

$$
\xrightarrow[\substack{\text{EtOH–H}_2\text{O}(9:1) \\ 20 \text{ min, } 60\%}]{h\nu, \text{ air}}
\tag{6.17}
$$

Photooxidation of phenols is of interest in environmental chemistry (41). Phenols have been found to exist in all form of natural water. It has been suggested that singlet oxygen is involved in such oxidation process.

Another useful oxidative reaction in aqueous media is the cleavage of cycloketones by hydrogen peroxide in the presence of Fe(II) salts [Eq. (6.18)]. The reaction proceeds through an α-hydroxy hydroperox-

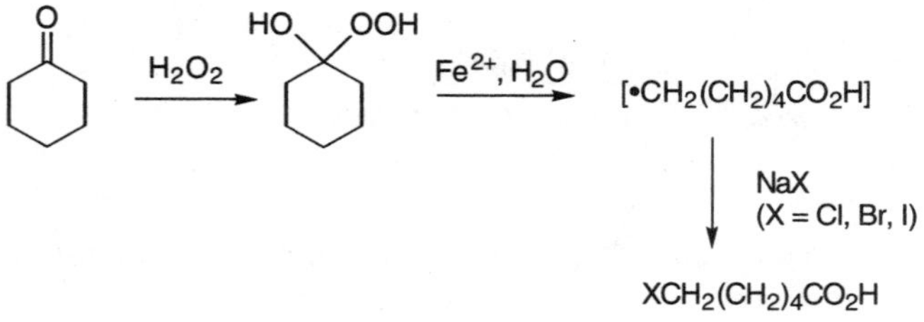

$$
\xrightarrow{\text{H}_2\text{O}_2} \xrightarrow{\text{Fe}^{2+}, \text{H}_2\text{O}} [\cdot\text{CH}_2(\text{CH}_2)_4\text{CO}_2\text{H}]
$$

$$
\downarrow \substack{\text{NaX} \\ (\text{X = Cl, Br, I})}
\tag{6.18}
$$

$$
\text{XCH}_2(\text{CH}_2)_4\text{CO}_2\text{H}
$$

ide, leading to a variety of products (42). The presence of Fe(II) salts decomposes the intermediate, generating a radical. In the presence of halide ions, the radical leads to synthetically useful ω-halocarboxylic acids (43).

6.4 REDUCTION

The reduction of carbonyl compounds by a number of reducing reagents has been carried out in aqueous media. Among these reagents, sodium borohydride is most frequently used. The reduction of carbonyl compounds by sodium borohydride can also use phase-transfer catalysts (44) or inverse phase-transfer catalysts (45) in a two-phase medium in the presence of surfactants. Other reagents for the reduction in aqueous media include a cadmium chloride–magnesium–THF–water system [Eq. (6.19)] (46), samarium iodide in aqueous THF (47), sodium

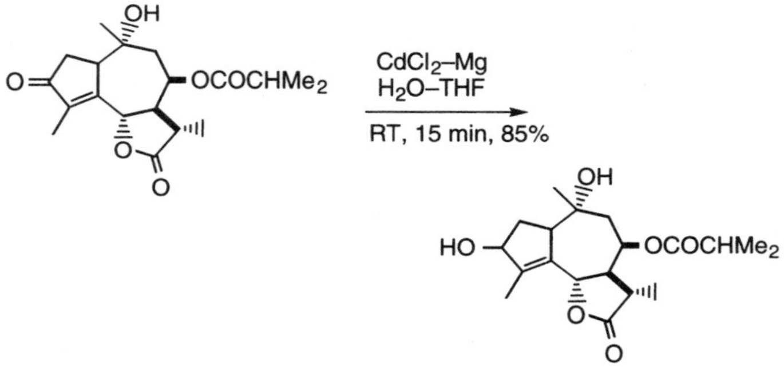

(6.19)

dithionite in aqueous DMF (48), sodium sulfide in the presence of poly-ethylene glycol (49), and metallic zinc together with nickel chloride (50). When the condition is controlled properly, the latter system is also very effective in reducing the C—C double bond of α,β-unsaturated carbonyl compounds [Eq. (6.4)]. The use of ultrasonication enhances the rate of the reaction (51).

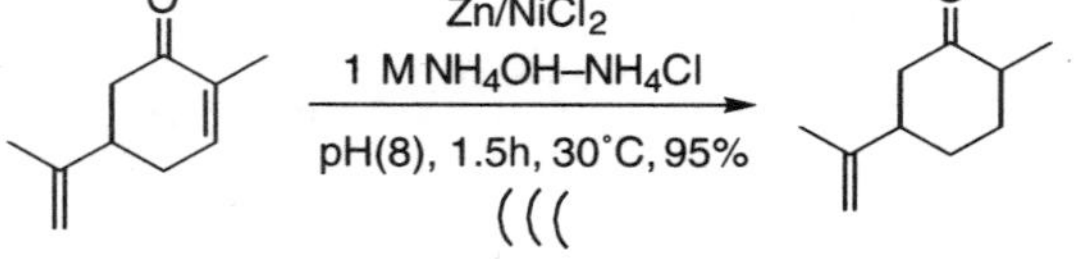

(6.20)

Sodium hydrogen telluride [(NaTeH)], prepared *in situ* from the reaction of tellurium powder with an aqueous ethanol solution of sodium borohydride, is an effective reducing reagent for many functionalities, such as azide, sulfoxide, disulfide, activated $C=C$ bonds, and nitroxide. Water is a convenient solvent for these transformations (52). A variety of functional groups, including aldehydes and ketones, olefins, and nitroxides, as well as azides, are also reduced by sodium hypophosphite buffer solution (53).

A number of reagents, such as sodium dithionite (54), zinc (55), chromous sulfate (56), and sodium iodide (57), have been used for the reductive removal of halogen from α-halo carbonyl compounds in aqueous medium. Aqueous chromous sulfate has also been used for the reduction of electron-deficient double bond (58). By using aqueous ethanol as solvent, 2,3-epoxyalkyl halides are readily transformed into allylic alcohols in the presence of a zinc–copper couple under ultrasonication (59):

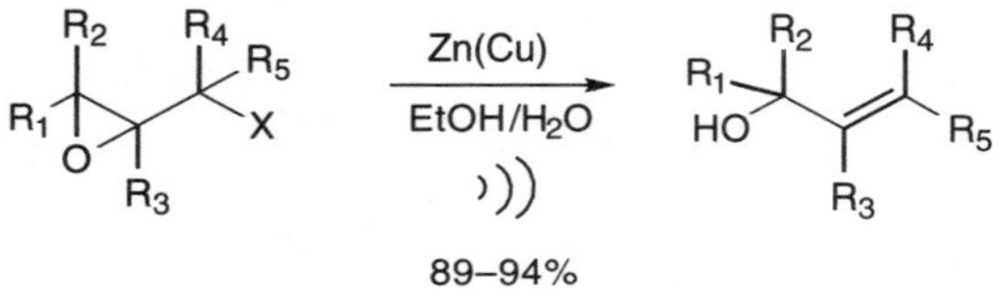

$$(6.21)$$

89–94%

Sulfones are versatile functionalities in organic synthesis (60). They can be easily removed when not needed. Vinyl sulfones have been stereospecifically reduced to the corresponding olefins with sodium dithionite in an aqueous medium (61):

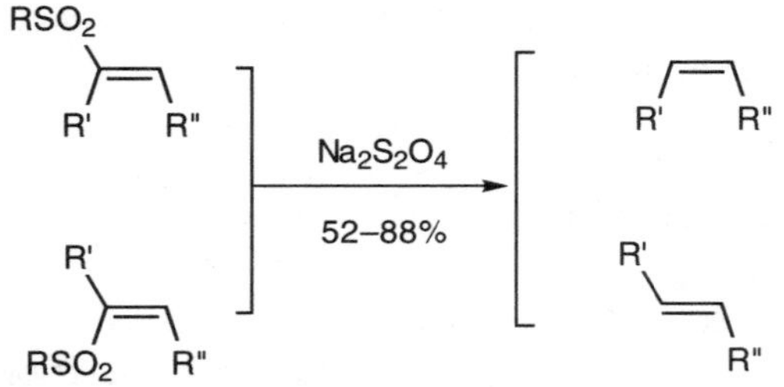

$$(6.22)$$

52–88%

The reduction of organic disulfide to dithiol is of interest in both chemistry and biology. The interchange between thiol and disulfide is important in the folding of proteins (62). A variety of reagents have been used to transform disulfides into thiols. For example, diaryl and dialkyl sulfide can be readily reduced by triphenylphosphine in aqueous organic solvents [Eq. (6.23)] (63). Whitesides, on the other hand, examined a number of water-soluble dithiols for reducing disulfide (64).

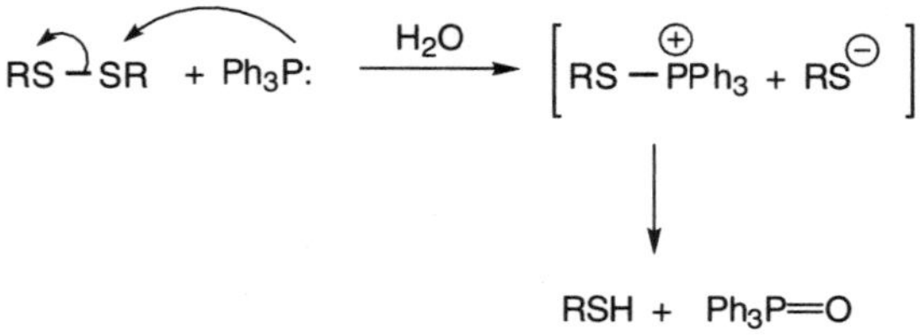

$$RS \overset{\frown}{-} SR \ + \ Ph_3P: \quad \xrightarrow{\ H_2O\ } \quad \left[RS \overset{\oplus}{-} PPh_3 \ + \ RS^{\ominus} \right]$$

$$\downarrow$$ (6.23)

$$RSH \ + \ Ph_3P{=}O$$

A water-soluble tin hydride (**7**) was recently prepared by Breslow (65). The tin hydride has three hydrophilic methyoxyethoxypropyl groups. In the presence of light or AIBN, it reduces a variety of alkyl halides in water. A noteworthy example of the reduction is the reaction with a water-soluble sugar derivative. In water, the bromine was cleanly removed:

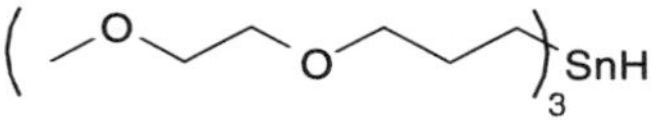

7

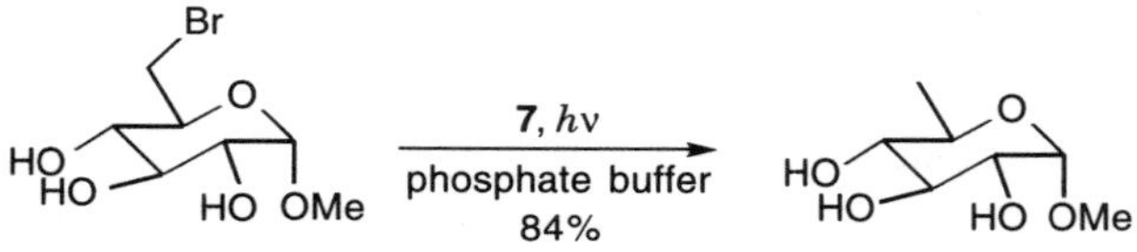

(6.24)

In studying the stereochemistry of radical reactions, it was found that limited amounts of water did not lead to diminished yield or stereoselectivity in the Lewis acid-mediated intermolecular radical addition to *N*-enoyloxazolidinones (66).

6.5 ELECTROCHEMISTRY

The electrochemical oxidation of carboxylic acids in forming decarboxylation coupling product, the so-called Kolbe reaction (Scheme 6.1) (67), is among the earliest reactions for organic synthesis. The initial

$$2\ RCO_2^- \quad \xrightarrow{\ -2e\ } \quad 2\ R^{\bullet} \ + \ 2CO_2$$

$$2\ R^{\bullet} \quad \longrightarrow \quad R{-}R$$

Scheme 6.1   The Kolbe reaction.

observation dates back to the 1830s by Michael Faraday (68). By now, electrochemical synthesis has become an independent discipline with a broad range of organic reactions. For performing electrochemical reactions, the conductivity of the reaction medium is an essential requirement. Thus water is one of the solvents most often used for electroorganic synthesis. However, recent studies have been focused on the use of organic solvents for such purposes, with respect to a higher selectivity and a higher operation concentration. Nevertheless, there is a distinctive advantage in using aqueous solution over organic solvents for electrochemical synthesis: the stability of the electrolysis medium (69). The oxidation–reduction of water yields O_2/H^+ and H_2/OH^-; whereas the oxidation–reduction of organic solvents commonly leads to a complex mixture of products that build up in the electrolyte and also leads to the loss of expansive solvent. Thus, electrolysis in water can be performed at a maintained level of pH without contaminating the electrolytic system.

Generally, electrochemical synthesis can be classified as anodic oxidative processes and cathodic reductive processes. Under anodic oxidative conditions, various organic compounds are oxidized. The product of such oxidations is influenced by the solvent, pH value of the medium, and the oxidative potential.

On the other hand, the cathode of electrolysis provides an electron source for the reduction of organic compounds. Electroreduction of unsaturated compounds in water or aqueous–organic mixtures generated products that are equivalent to catalytic hydrogenations. Generally, the rate of the reduction increases with the acidity of the medium (70).

Presently, electrochemical reactions are almost as diverse as nonelectrochemical reactions. In particular, the combination of electrochemistry with catalysts (electrochemical catalytic processes) (71), and with photochemistry (photoelectrochemical processes) (72) and enzymatic chemistry (electroenzymatic reactions) (73) make the scope of electrochemical reactions even broader. Most nonelectrochemical oxidation–reduction reactions have their counterparts in electrochemical processes. This book does not discuss this area in detail. Specialized monographs in electrochemistry are available for more information (74).

6.6 ENZYMATIC REACTIONS

Enzymes are highly efficient and highly selective biological catalysts that accelerate chemical reactions approaching the equilibrium. Reactions catalyzed by enzyme are 10^3–10^{17} times faster than the corresponding uncatalyzed ones. Reactions catalyzed by enzymes generally have high reaction specificity, stereospecificity, and substrate specificity. Because of these properties, the potential usefulness of enzymes for organic synthesis is becoming increasingly recognized. Water is the common medium for most enzymatic reactions. This book does not cover these reactions because many specialized treatises in this area are available in the literature (75). On the other hand, recent progress has been made by carrying out enzymatic reactions in organic solvents (76). The role of water has also been recognized in constructing aquasome as a possible vehicle for stabilizing biomolecules and for drug deliveries (77).

REFERENCES

1. Meunier, B. *Chem. Rev.* **92**, 1411 (1992).

2. Bernadou, J.; Fabiano, A. S.; Robert, A.; Meunier, B. *J. Am. Chem. Soc.* **116**, 9375 (1994).

3. Fringuelli, F.; Germani, R.; Pizzo, F.; Savelli, G. *Tetrahedron Lett.* **30**, 1427 (1989).

4. Fringuelli, F.; Germani, R.; Pizzo, F.; Santinelli, F.; Savelli, G. *J. Org. Chem.* **57**, 1198 (1992).

5. Berti, G. in *Topics in Stereochemistry*, Allinger, N. L. and Eliel, E. L., eds. Vol. 7, p. 93, John Wiley, 1967.

6. Kirshenbaum, K. S.; Sharpless, K. B. *J. Org. Chem.* **50**, 1979 (1985).

7. Gunstone, F. D. *Adv. Org. Chem.* **1**, 103 (1960).

8. Hofmann, K. A. *Ber.* **45**, 3329 (1912); Grieco, P. A.; Ohfune, Y.; Yokoyama, Y.; Owens, W. *J. Am. Chem. Soc.* **101**, 4749 (1979).

9. Milas, N. A.; Sussman, S. *J. Am. Chem. Soc.* **58**, 1302 (1936); Daniels, R.; Fischer, J. L. *J. Org. Chem.* **28**, 320 (1963).

10. Mugdan, M.; Young, D. P. *J. Chem. Soc.* 2988 (1949).

11. Sharpless, K. B.; Akashi, K. *J. Am. Chem. Soc.* **98**, 1986 (1976).

12. Van Rheenen, V.; Kelly, R. C.; Cha, D. F. *Tetrahedron Lett.* 1973 (1976).

13. D. G. Lee, *The Oxidation of Organic Compounds by Permanganate Ion and Hexavalent Chromium*, Open Court, La Salle, IL, 1980.

14. Mailhot, G.; Pilichowski, J. F.; Bolte, M. *New J. Chem.* **19**, 91 (1995).

15. Minato, M.; Yamamoto, K.; Tsuji, J. *J. Org. Chem.* **55**, 766 (1990).

16. Mugdan, M. Young, D. P. *J. Chem. Soc.* 2988 (1949).

17. For reviews, see Kolb, H. C.; Van Nieuwenhze, M. S.; Sharpless, K. B. *Chem. Rev.* **94**, 2483 (1994).

18. Hanessian, S.; Meffre, P.; Girad, M.; Beaudoin, S.; Sanceau, J. Y.; Bennani, Y. L. *J. Org. Chem.* **58**, 1991 (1993); Corey, E. J.; Jardine, P. D.; Virgil, S.; Yuen, P. W.; Connell, R. D. *J. Am. Chem. Soc.* **111**, 9243 (1989); Tomioka, K.; Nakajima, M.; Koga, K. *J. Am. Chem. Soc.* **109**, 6213 (1987); Tokles, M.; Snyder, J. K. *Tetrahedron Lett.* **27**, 3951 (1986); Yamada, T.; Narasaka, K. *Chem. Lett.* 131 (1986).

19. Hentges, S. G.; Sharpless, K. B. *J. Am. Chem. Soc.* **102**, 4263 (1980); Oishi, T.; Hirama, M. *Tetrahedron Lett.* **33**, 639 (1992); Imada, Y.; Saito, T.; Kawakami, T.; Murahashi, S. I. *Tetrahedron Lett.* **33**, 5081 (1992).

20. Kwong, H. L.; Sorato, C.; Ogino, Y.; Chen, H.; Sharpless, K. B. *Tetrahedron Lett.* **31**, 2999 (1990).

21. Gobel, T.; Sharpless, K. B. *Angew. Chem., Int. Ed. Engl.* **32**, 1329 (1993).

22. Crispino, G. A.; Jeong, K. S.; Kolb, H. C.; Wang, Z. M.; Xu, D.; Sharpless, K. B. *J. Org. Chem.* **58**, 844 (1993).

23. Kolb, H. C.; Andersson, P. C.; Sharpless, K. B. *J. Am. Chem. Soc.* **116**, 1278 (1994).

24. Hale, K. J.; Manaviazar, S.; Peak, S. A. *Tetrahedron Lett.* **35**, 425 (1994).

25. Abedel-Rahman, H.; Adams, J. P.; Boyes, A. L.; Kelly, M. J.; Mansfield, D. J.; Procopiou, P. A.; Roberts, S. M.; Slee, D. H.; Sidebottom, P. J.; Sik, V.; Watson, N. S. *J. Chem. Soc., Chem. Commun.* 1841 (1993).

26. Sharpless, K. B.; Chong, A. O.; Oshima, K. *J. Org. Chem.* **41**, 177 (1976).

27. Bäckvall, J. E. *Tetrahedron Lett.* **26**, 2225 (1975).

28. Hentges, S. G.; Sharpless, K. B. *J. Am. Chem. Soc.* **102**, 4263 (1980); Rubinstein, H.; Svendsen, J. S. *Acta Chem. Scand.* **48**, 439 (1994).

29. Li, G.; Chang, H. T.; Sharpless, K. B. *Angew. Chem., Int. Ed. Engl.* **35**, 451 (1996).

30. Trahanovsky, W. S.; Young, L. B. *J. Org. Chem.* **31**, 2033 (1966).

31. Ho, T. L. *Synth. Commun.* **9**, 237 (1979).

32. Balasubramanian, V.; Robinson, C. H. *Tetrahedron Lett.* **22**, 501 (1981).

33. Molander, G. A. *Chem. Rev.* **92**, 29 (1992).

34. Trost, B. M.; Curran, D. P. *Tetrahedron Lett.* **22**, 1287 (1981).

35. Zheng, T. C.; Richardson, D. E. *Tetrahedron Lett.* **36**, 833 (1995).

36. Nabjohn, N. *Org. React.* **24**, 261 (1976).

37. Leonard, N. J.; Hay, A. S.; Fulmer, R. W.; Gash, V. W. *J. Am. Chem. Soc.* **77**, 439 (1955).

38. Butler, R. N. in *Synthetic Reagents*, Pizey, J. S. ed., Vol. 4., Ellis Horwood, Chichester (U.K.), 1981.

39. Palou, J. *Chem. Soc. Rev.* **23**, 357 (1994).

40. Tateba, H.; Mihara, S. *Agric. Biol. Chem.* **55**, 873 (1991).

41. For examples, see Tratnyek, P. G.; Hoigne, J. *J. Photochem. Photobiol. A: Chem.* **84**, 153 (1994) and references cited theirein.

42. For a recent review on the reaction of α-hydroxy hydroperoxide, see Ganeshpure, P. A.; Adam, W. *Synthesis* 179 (1996).

43. Minisci, F. *Gazz. Chim. Ital.* **89**, 1910 (1959).

44. Lamaty, G.; Riviere, M. H.; Roque, J. P. *Bull. Soc. Chim. Fr.* 33 (1983).

45. Boyer, B.; Betzer, J. F.; Lamaty, G.; Leydet, A.; Roque, J. P. *New J. Chem.* **19**, 807 (1995).

46. Bordolol, M. *Tetrahedron Lett.* **34**, 1681 (1993).

47. Singh, A. K.; Bakshi, R. K.; Corey, E. J. *J. Am. Chem. Soc.* **109**, 6187 (1987); Hasegawa, E.; Curran, D. P. *J. Org. Chem.* **58**, 5008 (1993).

48. Krapcho, A. P.; Seidman, D. A. *Tetrahedron Lett.* **22**, 179 (1981).

49. Satagopan, V.; Chandalia, S. B. *Synth. Commun.* **19**, 1217 (1989).

50. Petrier, C.; Luche, J. L. *Tetrahedron Lett.* **33**, 5417 (1987); Baruah, R. N. *Tetrahedron Lett.* **33**, 5417 (1992).

51. Petrier, C.; Luche, J. L.; Lavaitte, S.; Morat, C. *J. Org. Chem.* **54**, 5313 (1989).

52. Petragnani, N.; Comasseto, J. V. *Synthesis* 1 (1986).

53. Boyer, S. K.; Bach, J.; McKenna, J.; Jagdmann, E. Jr. *J. Org. Chem.* **50**, 3409 (1985).

54. Chung, S.; K.; Hu, Q. Y. *Synth. Commun.* **12**, 261 (1982).

55. Rizzo, C. J.; Dunlap, N. K.; Smith, A. B. III. *J. Org. Chem.* **52**, 5280 (1987).

56. Castro, C. E.; Kray, W. C. Jr. *J. Am. Chem. Soc.* **85**, 2768 (1963).

57. Ono, A.; Fujimoto, E.; Ueno, M. *Synthesis* 570 (1986).

58. Castro, C. E.; Stephens, R. D.; Moje, S. *J. Am. Chem. Soc.* **88**, 4964 (1966).

59. Sarandeses, L. A.; Mourino, A.; Luche, J. L. *J. Chem. Soc., Chem. Commun.* 818 (1991).

60. Simpkins, N. S. *Sulphones in Organic Synthesis*, Pergamon Press, New York, 1993.

61. Bremner, J.; Julia, M.; Launay, M.; Stacino, J. P. *Tetrahedron Lett.* **23**, 3265 (1982).

62. Jocelyn, P. C. *Biochemistry of the SH Group*, Academic Press, London, 1972.

63. Schonberg, A.; Barakat, M. Z. *J. Chem. Soc.* 892 (1989); Overman, L. E.; O'Connor, E. M. *J. Am. Chem. Soc.* **98**, 771 (1976); Salim, A.; Tillett, J. G. *Phosph. Sulf. Silic.* **60**, 215 (1991).

64. Lamoureux, G. V.; Whitesides, G. M. *J. Org. Chem.* **58**, 633 (1993); Lees, W. J.; Singh, R.; Whitesides, G. M. *J. Org. Chem.* **56**, 7328 (1991).

65. Light, J.; Breslow, R. *Tetrahedron Lett.* **31**, 2957 (1990).

66. Sibi, M. P.; Jasperse, C. P.; Ji, J. *J. Am. Chem. Soc.* **117**, 10779 (1995).

67. Kolbe, H. *Ann. Chim.* **69**, 279 (1849).

68. Faraday, M. *Pogg. Ann.* **33**, 438 (1834).

69. Bersier, P. M.; Carlsson, L.; Bersier, J. in *Topics in Current Chemistry*, Vol. 70, Bunitz, J. D.; Hafner, S.; Ito, J. M.; Lehn, J. M.; Raymond, K. N.; Rees, C. W.; Vógtle, J. T. F., eds., Springer-Verlag, 1994.

70. Frumkin, A. N. in *Advances in Electrochemistry and Electrochemical Engineering*, Vol. 3, Delahay, P.; Tobias, C. W., eds. Interscience, New York, 1963.

71. *Topics in Current Chemistry*, Vol. 142, Steckhan, E., ed., Springer-Verlag, Berlin, 1987.

72. Fox, M. A. in *Topics in Current Chemistry*, Vol. 142, Steckhan, E. ed., Springer-Verlag, Berlin, 1987.

73. Steckhan, E. in *Topics in Current Chemistry*, Vol. 170, Steckhan, E. ed., Springer-Verlag, Berlin, 1994.

74. Kyriacou, D. K. *Basics of Electroorganic Synthesis*, Wiley, New York, 1981; Lund, H; Baizer, M. M. eds., *Organic Electrochemistry*, Marcel Dekker, New York, 1991.

75. For basic concepts and reviews, see Horton, H. R.; Moran, L. A.; Ochs, R. S.; Rawn, J. D.; Scrimgeour, K. G. *Principles of Biochemistry*, 2nd ed., Prentice-Hall, Englewood Cliffs, NJ, 1996; Whitesides, G. M.; Wong, C. H. *Angew. Chem., Int. Ed. Engl.* **24**, 617 (1985); Wong, C. H.; Halcomb, R. L.; Ichikawa, Y.; Kajimoto, T. *Angew. Chem., Int. Ed. Engl.* **34**, 412 (1995).

76. Klibanov, A. M. *Acc. Chem. Res.* **23**, 114 (1990).

77. Kossovsky, N. *Chemistry in Britain* **32**(2), 43 (1996).

CHAPTER 7

INDUSTRIAL APPLICATIONS

What you believe to be the summit is only a step further.
—Epikur (342–270 B.C.)*

At present, although there is a growing environmental awareness among the general public and industry, the major reason for the use of aqueous media in some large manufacturing processes is probably due to economic, rather than environmental, concerns. Listed in order of priority, these concerns are generally as follows:

- The major concern in governing most industrial processes is the economic value of the process. In this regard, water is the cheapest solvent available. Thus the use of water as solvent has a definite economic advantage over the use of more expensive organic solvents.
- The second concern in most industrial processes probably is the convenience of product separation. By carrying out reactions in water, one can isolate insoluble products by simple phase separation from the aqueous solution.
- The third concern probably is waste disposal. Again, to minimize the waste generation in chemical manufacturing is more of a cost than environmental issue. By proper design, an

*Quoted by Wolfgang F. Hoelderich of BASF AG, Gordon Conference on Environmentally Benign Organic Synthesis, 1996.

aqueous process can be recycled easily and used continuously, minimizing the disposal of waste.

• The fourth concern probably is environmentally related. Even though this is not an issue in many countries at the present stage, there is a growing interest in this regard. Regulations on environmental compliance could become increasingly tight in the future.

• The last advantage of using water relates to safety. Many organic solvents are imflammable and toxic, whereas water is not.

However, to draw a line between laboratory reactions and industrial processes is difficult except for the scale of operation. In principle, every laboratory reaction could be used in industry. Rather than review the use of aqueous medium for industrial processes in detail, this chapter offers some representative examples of successful industrial processes in aqueous media.

7.1 SYNTHESIS OF ADIPONITRILE

The production of adiponitrile is an important industrial process involving the electrohydrodimerization (EHD) of acrylonitrile. Adiponitrile is used as an important precursor for hexamethylenediamine and adipic acid, the monomers required for the manufacture of Nylon-66 polymer. The annual production of adiponitrile is about a million tons (1). Although it was initially studied in the 1940s, the electroreductive coupling of acrylonitrile to adiponitrile was commercialized only more than a decade later after Baizer (at Monsanto) developed the supporting electrolyte. It was found that a 90% yield of adiponitrile could be achieved when a concentrated solution of certain quaternary ammonium salts (QASs), such as tetraethylammonium–p-toluenesulfonate, is used together with lead or mercury cathodes (2):

$$2 \quad \diagup\!\!\diagdown\text{CN} \ + \ 2\,H_2O \quad \xrightarrow{\ 2e^-\ } \quad NC\diagdown\!\diagup\!\diagdown\!\diagup_{CN} \ + 2\,{}^-OH \qquad (7.1)$$

Initially, Monsanto employed a divided-cell EHD process, which was soon replaced by an undivided-cell process because of several shortcomings with the former process (3). The undivided-cell system involves electrolysis of a dilute solution of acrylonitrile in a mixed sodium phosphate–borate electrolyte using a cadmium cathode and a carbon steel anode. The presence of a quaternary ammonium salt is essential for the adiponitrile selectivity.

7.2 ASAHI'S SEBACIC ACID PROCESS

Sebacic acid is an important intermediate in the manufacture of polyamide resins, which have large worldwide demand. Electrochemical method has been explored by several companies, including BASF (4), Asahi Chemical Company (5), and a chemical plant in Russia (6), to replace the conventional method of alkaline hydrolysis of castor oil. In particular, the Asahi process has reported product yields as high as 92% with a 85–90% current efficiency. The electrochemical process involves three steps (Scheme 7.1), in which the key step is based on the

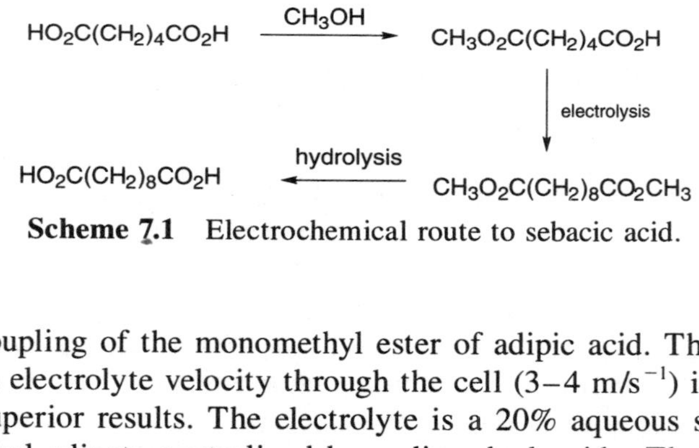

$$HO_2C(CH_2)_4CO_2H \xrightarrow{\ CH_3OH\ } CH_3O_2C(CH_2)_4CO_2H$$

$$\downarrow \text{electrolysis}$$

$$HO_2C(CH_2)_8CO_2H \xleftarrow{\ \text{hydrolysis}\ } CH_3O_2C(CH_2)_8CO_2CH_3$$

Scheme 7.1   Electrochemical route to sebacic acid.

anodic coupling of the monomethyl ester of adipic acid. The use of a very high electrolyte velocity through the cell (3–4 m/s^{-1}) is essential for the superior results. The electrolyte is a 20% aqueous solution of monomethyl adipate neutralized by sodium hydroxide. The operating temperature is 55°C with a platinum-plated titanium anode and a steel cathode. The cells are generally operated at a current density of 20 A/dm^2 and 14 V.

7.3 RUHRCHEMIE/RHÔNE-POULENC'S HYDROFORMYLATION PROCESS

The hydroformylation process is one of the most successful applications of aqueous-medium catalysis in industrial manufacture. It started in 1982 with a series of patents, including the synthesis of aldehydes (7), the recovery of rhodium catalyst (8), and the preparation of water-soluble sulfonated phosphane ligands (9). During this time, production of   3,3′,3″-phosphinidynetris(benzenesulfonic   acid)   trisodium   salt (tppts) (**1**) at the ton scale was realized (10). The Ruhrchemie/Rhône-

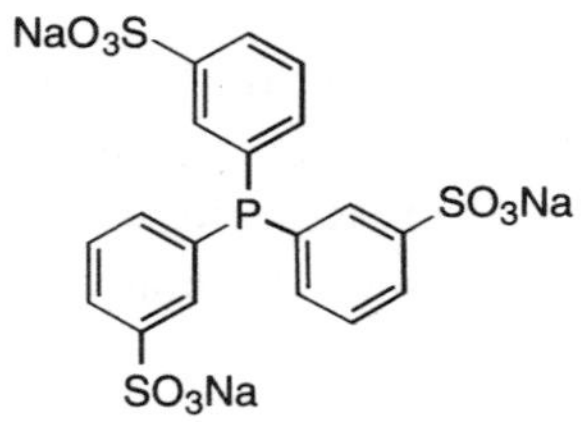

1

Poulenc process went into operation 2 years later. The catalyst used is $HRh(CO)(tppts)_3$. The initial capacity was 100,000 tons per year, and the present annual production of butyraldehyde alone exceeds 0.25 million tons. The product is separated from the catalyst solution by a simple phase separation, and the catalyst solution is recharged to the reactor for further reaction. During the process, the loss of rhodium catalyst in the organic phase is negligible. The reaction provides predominantly *n*-butyraldehyde [Eq. (7.2)]. Alcohols have been synthe-

$$CH_3{-}CH{=}CH_2 \ + \ CO \ + \ H_2 \quad \xrightarrow[\text{H}_2\text{O}]{\text{cat. HRh(CO)(tppts)}_3} \quad CH_3CH_2CH_2CHO \quad (7.2)$$

sized by treating an olefinic hydrocarbon with carbon monoxide and hydrogen in a hydroformylation zone using a rhodium complex catalyst and an amine modifier to effect the reaction. After the formation of the alcohol, the catalyst is extracted from the alcohol by treatment with an aqueous ammonium hydroxide solution. Then, the aqueous ammonium solution containing the rhodium catalyst is subjected to an extraction process utilizing the amine modifier as the extractant and thereafter recycling the rhodium complex catalyst and amine modifier to the hydroformylation zone (11).

7.4 WACKER–CHEMIE'S OXIDATION PROCESS

Two different processes for the Wacker oxidation have been used: the one-stage process, which olefin and oxygen are reacted simultaneously with the catalyst solution; and the two-stage process, in which the olefin is first allowed to react with the catalyst solution, then the carbonyl compound formed is separated and finally the reduced catalyst solution is oxidized by oxygen or air (12). The use of water as solvent provides the basis for separation of the product and recycling of the catalyst.

Through this process, ethylene is more than 99% converted to acetaldehyde when passed (together with the catalyst solution) once through a flow reactor at moderate pressure and medium temperatures [Eq. (7.3)]. The acetaldehyde formed is removed by distillation. The

$$CH_2{=}CH_2 + O_2 \xrightarrow[\text{H}_2\text{O}]{\text{Pd(II)/Cu(II)}} CH_3CHO \qquad (7.3)$$

catalyst solution from which the aldehyde has been removed is then transferred back, with a circulating pump, into the oxidation reactor. After reacting with air, it flows back to the ethylene reactor. The crude product aldehyde is purified in a two-stage distillation. Lower-boiling byproducts such as CH_3Cl, CH_2Cl_2, and CO_2 are separated first, followed by removal of higher-boiling byproducts such as chlorinated acetaldehydes, chloroform, methylene chloride, ethylene chloride, ethylene chlorohydrin, and acetic acid, as well as water.

7.5 SYNTHESIS OF FURFURAL FROM BIOMASS

There has been considerable interest in the conversion of cellulosic biomass into useful materials (13). Among the methods being developed, the Hoppe–Seyler process converts biomass into oil, gas, water-soluble organic compounds, and other substances using water and alkaline (14). Alternatively, transformation of biomass to useful compounds uses aqueous acidic conditions. One example is the synthesis of furfural. All aldopentoses form 2-furfuraldehyde in high yield when exposed to aqueous acidic solution at elevated temperature. The furfural formation is through a dehydration process. The generally accepted mechanism for the dehydration is illustrated in the case of D-xylose (Scheme 7.2) (15). Besides furfural formation, biomass has been transformed to a variety of other compounds in aqueous solution (15).

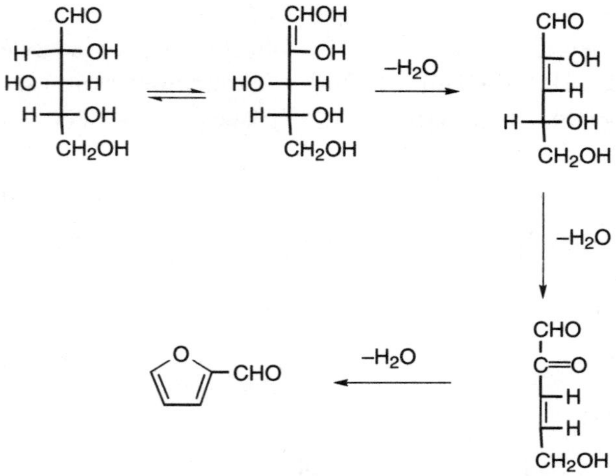

Scheme 7.2

7.6 OTHER PROCESSES

Various other aqueous processes are currently being commercialized or explored in industry. These processes include the selective hydrocyanation of butadiene catalyzed by Ni(0)/triarylphosphite complexes [Eq. (7.4)] (invented by Drinkard) (16), the hydrodimerization of butadiene using a biphasic system with palladium/tppts catalyst [Eq. (7.5)]

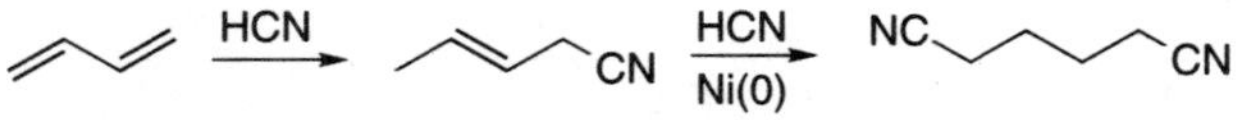

$$\text{(7.4)}$$

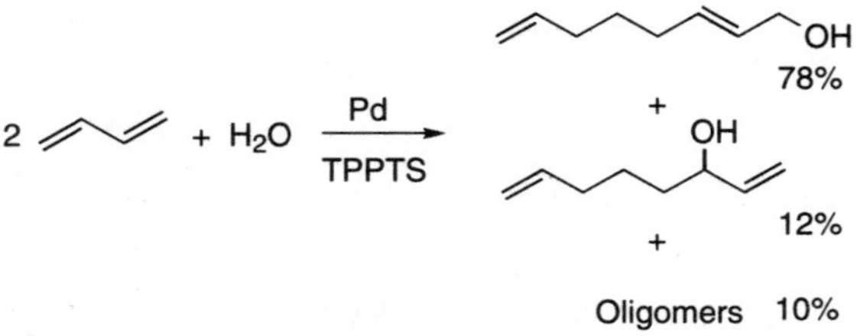

$$\text{(7.5)}$$

by the Kuraray Company for the production of 1,9-nonanediol (17), the Atlas Powder Company's electrochemical reduction of glucose for the manufacture of sorbitol and mannitol (18), the electrochemical reduction of phthalic acids to the corresponding dihydrophthalic acids (19), the electrochemical reductive coupling of acetone to yield pinacol (20), the electrochemical oxidation of 1,4-butynediol to acetylene dicarboxylic acid (21), the Otsuka Chemical Company's synthesis of maltol from furfural by the electrochemical oxidation (Scheme 7.3) (22), as

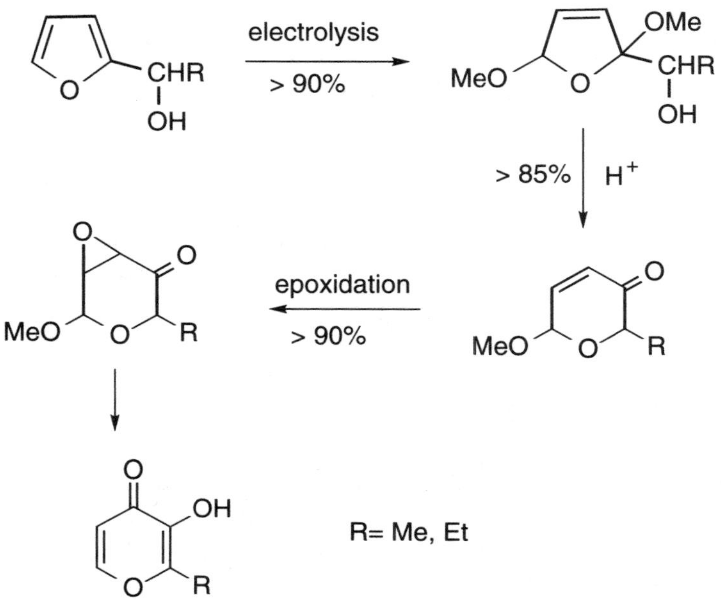

Scheme 7.3 Otsuka's maltol synthesis.

well as a variety of enzymatic methods in various biotech industrial processes (23). Tribenzyltin chloride and dibenzyltin dichloride has been prepared from benzyl chloride and tin in a mixture of water and benzene [Eq. (7.6)] (24).

$$2\ PhCH_2Cl\ +\ Sn\ \xrightarrow[\text{reflux}]{\text{H}_2\text{O/benzene}}\ (PhCH_2)_2SnCl_2 \qquad (7.6)$$

A process has been developed for alkylation at carbon and phosphorus sites in an aqueous medium using palladium catalysts containing sulfonated triarylphosphines. Aryl and olefinic halides are coupled with acetylenes, olefins, alkenyl or arylboric acid, and dialkyl phosphite in this way [Eqs. (7.7)–(7.9)] (25). When alkyne is used, a catalytic amount of a copper(I) salt, such as copper halide or copper nitrate, is required.

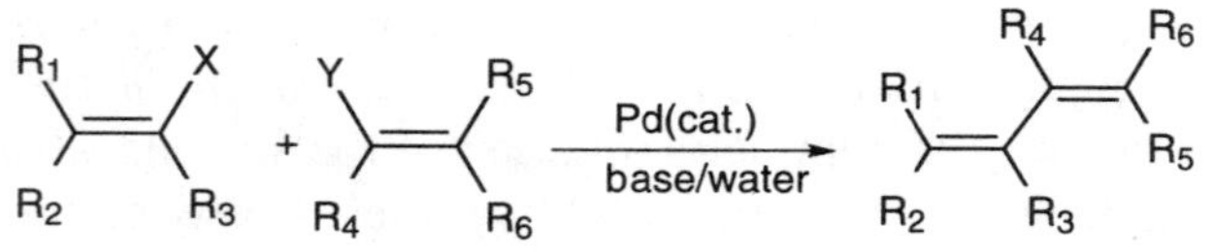

$$\text{(7.7)}$$

$$X = \text{halide}, \ Y = \text{H}, \ \text{B(OH)}_2$$

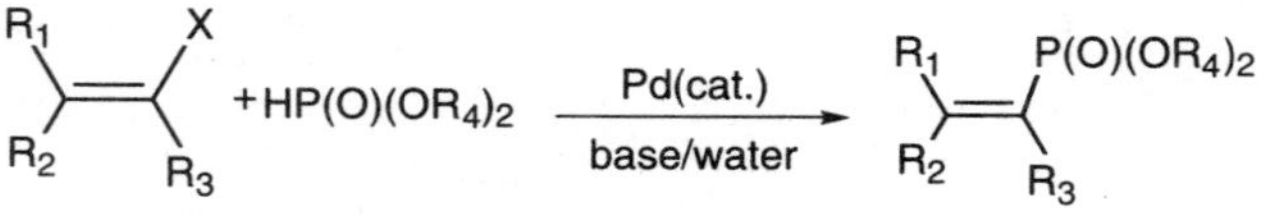

$$\text{(7.8)}$$

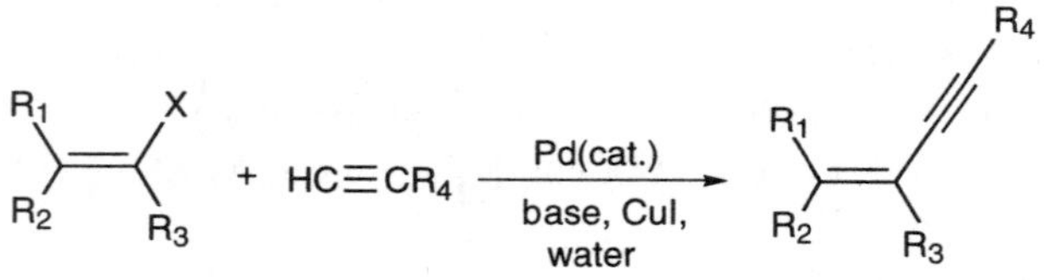

$$\text{(7.9)}$$

Diynoic acids are prepared from alkynoic acids and 1-haloalkynes in an aqueous solution in the presence of a catalytic amount of cuprous halide together with a reducing reagent, such as hydroylamine hydrochloride, and an alkylamine base (26):

$$CH_3-(CH_2)_m-C\equiv C-X \ + \ HC\equiv C-(CH_2)_n-CO_2H \ \xrightarrow[\substack{HONH_2 \bullet HCl \\ Et_3N/water}]{\text{cat. CuI}}$$

$$X = Cl, Br, I \qquad \text{(7.10)}$$

$$CH_3-(CH_2)_m-C\equiv C-C\equiv C-(CH_2)_n-CO_2H$$

The polymerization of 1,3-butadiene to give syndiotactic 1,2-polybutadiene has been carried out in an aqueous medium in the presence of a cobalt catalyst (27). The polymer prepared by this method has a melting point of <195°C. An electrochemical process has been developed for oxidizing aromatic and alkyl-substituted aromatic compounds to the corresponding quinones in an aqueous, weakly acidic solution (28). In this process, ceric ions are generated electrochemically.

7.7 CONCLUSIONS

Even with the many promises associated with the use of water as a solvent in industrial processes, it is not likely to be the panacea in preventing industrial pollution. Nevertheless, it is an important first step in the overall strategy toward reducing pollutant emission in the environment. Furthermore, the commercialization of the various advantageous aqueous methodologies for large-scale syntheses, especially catalysis with water-soluble catalysts, is expected to increase rapidly.

REFERENCES

1. Danly, D. E.; King, C. J. H. in *Organic Electrochemistry*, 3rd ed., Lund, H.; Baizer, M. M., eds., Marcel Dekker, New York, 1991.
2. Baizer, M. M.; Danly, D. E. *Chemtech.* **10**(3), 161 (1980).
3. Danly, D. E.; *AIChE Symp. Series* **77**(204), 39 (1981).
4. U.S. Patent 3,652,430 (1972), Beck, F.; Himmele, W.; Haufe, J.; Brunold, A. (to BASF AG).
5. U.S. Patent 3,896,011 (1975), Isoya, T.; Kakuta, R. Kawamura, C. (to Asahi Kasei).
6. Kovsman, E. P. *Sov. Chem. Ind.* **1**, 13 (1973).
7. DE-B 323470 (1982), Cornils, B.; Hibbel, J.; Konkol, W.; Lieder, B.; Much, J.; Schimd, V.; Wiebus, E. (to Ruhrchemie AG).
8. DE-B 3235029 (1982), Gärter, R.; Cornils, B.; Bexten, L.; Kupies, D (to Ruhrchemie AG).
9. DE-B 3235030 (1982), Gärter, R.; Cornils, B.; Springer, H.; Lappe, P. (to Ruhrchemie AG).
10. DE-B 3431643 (1984), Bexten, L.; Cornils, B.; Kupies, D. (to Ruhrchemie AG).
11. U.S. Patent 4,329,521 (1982), Homeier, E. H.; Imai, T.; Mackowiak, D. E.; Guzolek, C. E.
12. Smidt, J.; Hafner, W.; Jira, R.; Sieber, R.; Sedlmeier, J.; Sabel, A. *Angew. Chem., Int. Ed. Engl.* **1**, 80 (1962).
13. For a review, see Theander, O.; Nelson, D. A. *Adv. Carbohydr. Chem. Biochem.* **46**, 273 (1988).
14. Hoppe-Seyler, F *Ber.* **4**, 15 (1871).
15. Feather, M. S. *Tetrahedron Lett.* 4143 (1970); Isbell, H. S. *J. Res. Natl. Bur. Stand.* **32**, 45 (1944).
16. U.S. Patent 3.496.215 (1970), Drinkard, W. C.; Lindsay, R. V. (to DuPont).
17. Fr. Patent 2.366.237 (1976), Kuntz, E. (to Rhône-Poulenc); U.S. Patent 4808756 (1989), Tokitoh, Y.; Yoshimura, N. (to Kuraray Co.); EP-B 04362226 (1990), Yoshimura, N.; Tamura, M. (to Kuraray Co.).
18. Taylor, R. L. *Chem. Met. Eng.* **44**, 588 (1973).

19. Nohe, H. *Chem. Ing. Tech.* **46**, 594 (1974).

20. U.S. Patent 3,992,269 (1976), Sugano, T. T.; Schenber, B. A.; Walburg, J. A.; Shuster, N. (to Diamond Shamrock Corp.).

21. Degner, D. *Technique of Electroorganic Synthesis*, Part III, Weinberg, N. L.; Tilak, B. V. eds., Wiley, New York, 1982.

22. Taniquchi, M. in *Recent Advances in Electroorganic Synthesis*, Tori, S., ed., Elsevier, Amsterdam, 1987.

23. *Biocatalysts for Industry*, Dordick, J. S. ed., Plenum Press, New York, 1991.

24. Sisido, K.; Takeda, Y.; Kinugawa, Z. *J. Am. Chem. Soc.* **83**, 538 (1961); Sisido, K.; Kozima, S.; Hanada, T. *J. Organomet. Chem.* **9**, 99 (1967); Sisido, K.; Kozima, S. *J. Organomet. Chem.* **11**, 503 (1968).

25. U.S. Patent 5,043,510 (1991), Casalnuovo, A. L.; Nugent, Jr., W. A. (to DuPont).

26. U.S. Patent 4,867,917 (1989), Schnur, J. M.; Singh, A.

27. U.S. Patent 5,011,896 (1991), Bell, A. J.; Ofstead, E. A. (to Goodyear).

28. U.S. Patent 4,670,108 (1987), Kreh, R. P.; Spotnitz, R. M. (to W. R. Grace).

INDEX

Vinylphosphonium salts, 52
Volume of activation, 8, 13
  Claisen rearrangements, 39
  Diels–Alder reactions, 13
  nucleophilic additions, 50

Wacker oxidation, 150
Wacker–Chemie's oxidation
    process,
  one-stage process, 183
  two-stage process, 183
Water
  cohesive energy density (CED), 6
  density, 5
  internal pressure, 6
  physical properties, 4
  specific heat, 5
  surface tension, 5
  viscosity, 5
  three basic forms, 3
  phase diagram, 3
  structure, 3
Water-soluble tin hydride, 174
Water-soluble ligands, 116–123
Water-soluble metalloporphyrins,
    161

Water-soluble Pd catalyst,
    carbonylation, 132
Water-soluble Pd catalyst,
    coupling, 142
Water-soluble phosphines, 52,
    116–123
Water-soluble Rh catalyst, 126
Water-tolerant catalysts,
    Diels–Alder reaction, 26
Weitz–Scheffer epoxidation, 163
Wieland–Miescher ketone, 51
Wilkinson's catalyst, 124
Wittig olefination, 55
Wittig type reaction, 94
Wittig–Horner, 55
Wurtz coupling, 65, 97

Ytterbium triflate 51, 55, 105
Yohimbine, 31

Zeise's salt, 150
Zn(0)/BiCl$_3$, 179
Zn–Cu couple, 95, 97, 98, 173
Zn–InCl$_3$, 77

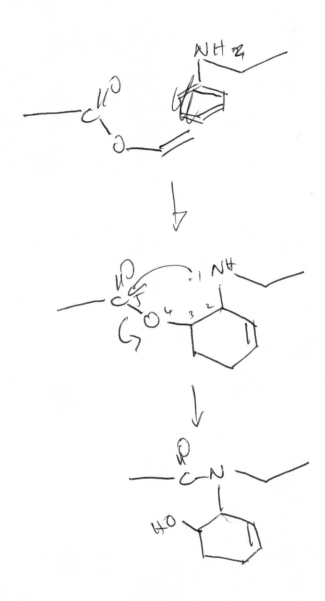

Insertion Rxn — in Peptide Chemistry